Globalization and the Race for Resources

Themes in Global Social Change

Series Editor
Christopher Chase-Dunn

Consulting Editors
Janet Lippman Abu-Lughod
Giovanni Arrighi
Jonathan Friedman
Keith Griffin

Globalization and the Race for Resources

Stephen G. Bunker and Paul S. Ciccantell

The Johns Hopkins University Press
Baltimore

© 2005 The Johns Hopkins University Press
All rights reserved. Published 2005
Printed in the United States of America on acid-free paper
9 8 7 6 5 4 3 2 1

The Johns Hopkins University Press
2715 North Charles Street
Baltimore, Maryland 21218-4363
www.press.jhu.edu

Library of Congress Cataloging-in-Publication Data

Bunker, Stephen G., 1944–
    Globalization and the race for resources / Stephen G. Bunker and
Paul S. Ciccantell.
        p. cm. — (Themes in global social change)
    Includes bibliographical references and index.
    ISBN 0-8018-8242-7 (hardcover : alk. paper)
    ISBN 0-8018-8243-5 (pbk. : alk. paper)
    1. Natural resources—History. 2. Mineral industries—History.
3. Globalization—History. 4. Capitalism—History. 5. International
economic relations—History. I. Ciccantell, Paul S., 1965– II. Title.
III. Series.
    HC85.B86 2006
    333.7—dc22   2005006240

A catalog record for this book is available from the British Library.

Additional reference materials for this book are available at
www.ssc.wisc.edu/globalization_and_the_race_for_resources/.

Contents

Figures and Tables

Figures

Tables

Preface: Finding the Global in the Local

The inspiration for this study of the forces that have driven the last six centuries of global trade expansion comes from a single, very local, and very remote place—a huge man-made hole deep in the Amazon jungle. Like the Congo River in Conrad's *Heart of Darkness*, the Amazon River provides fairly easy access, by boat, to the center of its own drainage basin, but traversing the vast landmass around it often involves penetrating mile after mile of dense vegetation and wet, broken ground. Carajás, nine hundred kilometers from the coast on the southern edge of the Amazon, was so hard to reach that only members of the small indigenous groups that lived in the area and the occasional explorer or adventurer had ever been there. By the beginning of the 1970s, though, authorizations and negotiations were moving forward on what would, by 1985, be the largest iron mine in the world. Eventually, airstrips, then a road, and finally a railroad and a port would be built, at great expense and with a single purpose—to send North America, Asia, and Europe enough iron to satisfy more than 10 percent of the world's increasingly voracious demand for steel.

The global market for iron that created the incentives and generated the finance for digging such a huge hole in such a remote place was at first incidental to what drew me to study Carajás. I wanted to see if Carajás contradicted or reinforced my earlier studies of natural resource extraction. Using the environmental and economic history of the nineteenth-century rubber boom in the Amazon and the ways it had contributed to the tremendous technical and financial advances in the industrial nations while leaving the Amazon desperately impoverished, I had argued that extractive economies had severely damaged the local environment, the various societies that lived there, and their prospects for future economic and political development (Bunker 1985). Critics of these pessimistic conclusions, though, claimed that twentieth-century extraction of mineral resources had far more beneficial effects than the earlier extraction of vegetable matter that I had analyzed (Katzman 1987). Carajás seemed to offer an ideal test of their claim.

I quickly discovered that while the potential environmental and social problems that the development of Carajás might bring deeply concerned national and local activists and intellectuals, most of the intense public discussion concerned how the mine would be financed, how its costs and benefits would be distributed, and what the size and location of the market it served would be. Underlying the debates about these issues, and contributing enormously to their intensity, was an undeclared struggle between United States Steel and a union of Japanese steel companies about where and in what quantity the outtake of the mine would be shipped, and thus over the rate of extraction, the size of the railroad and port, the scale of necessary investment and debt, and the sources of the financing. Given Japan's much greater distance from Carajás, and its much more intense raw material needs, the preferred solutions of Japan and the United States to these questions were intensely opposed. Both sides, however, controlled resources extremely important to the mine's success, so neither could be ignored.

Many of the Brazilian businessmen and politicians involved in these negotiations and decisions told me that dealing with the Japanese was particularly confusing, as they would appear to propose or agree to something at one meeting and then contradict themselves or deny their agreement, usually with no explanation, at the next. Each Japanese agent was clearly pursuing his own nation's or firm's interests, but all cast their arguments in terms of what would best serve Brazil's economy. They buttressed their arguments with complex economic logic and formulas, in some cases presented in glossy published studies they had conducted of how the plans they proposed would benefit the Amazonian and the Brazilian economies. U.S. and European interests did not cloak their proposals in the same complex theoretical language and mathematical formulas, but they were quite explicit about their preferences, claiming that their solutions would bring more benefits and cost less money than the Japanese proposals. The Japanese, the Europeans, and the Americans were all making strategic alliances with Brazilian firms and government agencies, whose directors they were able to convince about the benefits of their particular proposals.

These negotiations, and the obfuscating rhetoric with which they were carried out, were clearly going to have important effects on how the mine operated. I decided that in order to understand Carajás, I would need to understand first what the Japanese, the Europeans, and the Americans were proposing and then study why their proposals were so different.

I soon realized that what was at stake in Carajás was whether the United States or Japan would control the world steel industry, and how the outcome of that struggle would affect the European economy. I also realized that the contest was not limited to iron and steel; the Japanese and American firms were competing to participate in mining the bauxite that had been discovered in the Amazon, and then to use the cheap hydroelectric power that they hoped could be generated there to smelt the bauxite into aluminum. The Amazon was rapidly being transformed into a major supplier of some of the minerals most critical to global industry, and it was very clear that powerful and astute industrial interests would try to shape all of the decisions about extracting and exporting these minerals. As the scope of the study expanded, I needed help to keep track of all of the complexly interlinking parts. Paul Ciccantell first joined in the research to study how the world aluminum industry was affecting and affected by the Amazon and the different national interests struggling to control it; he then joined me in research on Japanese strategies to dominate trade through controlling access to raw materials.

We were coming to understand how important the U.S.-Japanese struggle for trade dominance would be in shaping decisions that would impact the ecological and social integrity of the Amazon, but we still felt it was tremendously important as well to analyze the local interaction between economy and ecology that the extraction of iron and bauxite would involve and to compare that interaction with the effects of earlier extractive economies there. As I looked back at the earlier trade in rubber, in cacao, in wood, and in turtle-egg oil, from the perspective of what I was learning about Carajás I realized that those earlier extractions were also shaped by national struggles for trade dominance, between the Portuguese and the Dutch in the sixteenth century, between the Dutch and the British in the seventeenth and eighteenth centuries, and between the British and the Americans in the nineteenth and twentieth centuries. I also saw, though, that in each extractive economy, the physical features of the natural resource itself, where and how it lay on or in the earth or water, the topography of the site from which it was extracted, and the trajectory of the rivers along which it could be transported—all directly determined both the economic organization and the ecological effects of its extraction and export. In other words, the local conditions that produced the resource directly shaped international strategies to export it.

As the world economy became more industrialized, the technologies used in local extraction became more efficient and more powerful, increasingly expensive, and were increasingly imported from the industrial centers. This was especially true of transport; the first extractive economies that incorporated the Amazon into world markets depended on small wooden boats powered by sails. Rudimentary oceangoing steamboats were just appearing as the rubber boom started two centuries later. These enabled the export of far greater volumes, at far lower prices per pound, than had been possible before. By the time Carajás opened in 1985, reinforced steel hulls could carry cargos almost a hundred times larger than had been possible during the rubber boom. The extra size and speed available so reduced the cost per unit of weight shipped that the Japanese could now transport iron ore halfway around the world more cheaply than U.S. Steel could ship its iron ore across the Great Lakes. These solutions perfectly fit Japan's need for an ever greater volume of raw materials to feed its rapidly growing and increasingly competitive production of industrial goods.

The history of Amazonian extraction thus shows how the development of shipping technology was a critical, and very concrete, element in the material intensification and spatial expansion of the world economy over the past six hundred years, proving that the nation that could develop and control the most efficient shipping could also guarantee its access to cheap raw materials. Paul and I soon realized that Japanese firms and states had collaborated to develop transport technologies in order to achieve national trade dominance. From that realization, the question that emerged was whether similar collaborations on transport technology had shaped earlier national campaigns to gain the cheapest and most stable access to the raw materials most voluminously used in industrial production—in other words, to achieve trade dominance. We set out to look at how the Dutch, the British, and the Americans had obtained privileged access to critical raw materials by developing the most efficient transport systems in each respective era. In the process, we saw that developing such transport systems generated whole new systems of firm-state collaboration and financial institutions powerful enough to handle much larger investments distributed over a much broader space. Without intending to, we found a historically consistent dynamic of globalization and a model to measure the growing inequality between nations that globalization entails.

Carajás, and the sequence of Amazonian extractive economies that preceded it, thus gave us a material and spatial perspective on globalization. It is one that directly contradicts those influential theoretical formulations that see globalization as (a) a recent and unique phenomenon that (b) deterritorializes production and (c) supersedes the national state. From the perspective of raw material access and transport, we see globalization as simply the most recent manifestation of capitalism's internal dynamic, which remains deeply rooted in place and involves the active participation of states. This essential continuity enables us to use concepts rooted in classical theories of political economy and economic history in our analysis of the processes that lead to globalization.

Capitalism is deeply rooted in the ongoing, cumulatively sequential expansion of its own reproduction. This expanded reproduction is achieved through the introduction of new technologies that enable expanded economies of scale. The economic viability of expanding economies of scale depends on accelerated throughput of matter and energy. Humans can create neither matter nor energy; they can only extract and transform them whence they are produced in nature. Natural production of the particular material and energetic forms that specific technologies require occurs only in particular places on the globe, at rates and in volumes that do not respond to human demand. The additional volumes and types of raw material that new technologies require can only be obtained from ever more distant places as proximate source are depleted. This increases the cost of raw materials, which undermines the economies of scale. Historically, capital has responded to this contradiction by increasing the size and speed of transport in ways that reduce the ton-mile cost of moving large volumes of raw material.

Because national states depend on the trade revenues and on the wages to labor of firms in their own territory, and because raw material transformation is critical to their military potential as well, national states have participated in industrial searches for raw materials. This dynamic allies firms, sectors, and states in those nations that successfully bid for world trade dominance.

In sum, economies of scale require increased volumes of increasingly cheap raw materials, which simultaneously make it possible and necessary for capital to intensify and expand material production. This in turn depends on expanded consumption of matter, which can only be sustained by exploitation of natural resources across ever broader spaces. Capitalism has always intensified materially in ways that required territorial and spatial expansion. In this sense, "globalization" is properly seen not as the supersession of classical relations of class, labor, and politics within a territorial base, but rather as a notice that a historically constant process of expansion may be reaching its global limits materially or ecologically.

Until the exhaustion of new materials and new spaces from which to extract them forces a change in the internal dynamic of capital, the absolutely immutable location of essential raw materials in particular places must maintain the fundamental territoriality of production, so questions of state sovereignty over territory and resources remain, and will remain, politically, economically, and environmentally crucial.

We will examine these processes first by analyzing the sequence of Amazonian extractive economies. For each such economy, we will explain how its ecological-economic interactions across local, national, and global spaces were affected by previous incorporations of Amazonian matter and space into the world economy, by the historical relations of Amazonia to the rest of Brazil and to the Brazilian state, and by the reiterated, scale-expanding, competition for trade dominance in the world economy.

We will then use what the Amazon teaches us about matter and space, and about society and nature, to explain the ways that different nations achieved trade dominance by creating technological, financial, and social instruments and institutions that provided them with cheaper and more stable access to raw materials than their competitors could manage. We will show how each nation that achieved trade dominance over a limited portion of the globe provided the technology and created the financial forms that enabled its successor to expand the world economy over a broader space. Then we will argue that huge, remote projects like Carajás demonstrate that this process has finally become global in the physical sense of covering the entire globe. This means that the old forms of competition for trade dominance, which have become increasingly destructive socially and environmentally, are now reaching their physical limits.

We will end by considering how the citizens and activists of the material-consuming core and of the resource-rich periphery can help each other respond to this impending crisis in the competition over the environment and search for means by which to preserve the social and ecological integrity of the nations that supply raw materials and, in the process, to change the materially voracious and environmentally destructive dynamic of the global economy.

I received a huge amount of help in the research and writing of this book. My first and deepest thanks go to Roberto Araújo Santos. I have tried to model my own efforts on his intellectual and spiritual example. Though my success in following his example has been partial at best, his influence has permeated all of my work on the Amazon and improved it greatly. Roberto and his wife, Betty, have also provided an abode during and a refuge from the rigors of fieldwork, together with wise counsel about its interpretation. It was in their living room and in Roberto's study in Belém that this work was conceived and its first chapters were written.

I would not have been able to complete this work as well and as quickly as I did without the extraordinarily warm and encouraging support and assistance that I received from the office staff of the Sociology Department at the University of Wisconsin in Madison. Toni Schulze, Patrick Brenzel, Molly Swank, and Molly's sister Katie cheerfully and patiently—and with great good humor, concern, and skill—provided exactly the assistance I needed to keep working through the pain, discouragement, fatigue, and occasional low energy brought on by the frequent, increasingly intrusive, and only partially successful medical attempts to cure what was ailing me. The various colleagues and friends who chaired the department during those years, Chuck Halaby, Adam Gamoran, and Pam Oliver, went to extraordinary lengths to keep me comfortable and productive during this time and tolerated the extra burdens that I was putting on an already very busy staff.

Dwight Haase has done yeoman work in the preparation of various figures and tables, many of which cannot be in this volume but will serve in other publications.

Gay Seidman, Jane Collins, and Erik O. Wright, wonderful friends and colleagues, gave me brilliant readings of the first full draft, and Jane has continued to comment beautifully on its revisions. I was blessed that my office was located between those of Gay and Erik, and I received huge support from both of them through this difficult time. Gay organized a wonderful conference to celebrate my contributions to sociology. She brought together graduate students I had trained over the years and close friends and colleagues with whom I had worked closely in different times and places. The papers presented were stellar, and the love and excitement in that group boosted the strength and energy I needed to keep working.

Jonathan Leitner expanded on my pleasure in directing his wonderful thesis and dissertation, and in reading his publications since graduating, by reading each chapter as I finished it and then re-reading it after I had incorporated his suggestions. Denis O'Hearn, Amanda Okopski, and Andrew Schrank all made very helpful comments on different parts of the manuscript. I only learned while writing this book, after knowing her well for almost sixty years, that my mother, Priscilla Fleitmann Bunker, is an absolutely brilliant reader and editor. I attempted to recover from my tardiness by sending her each chapter as I finished it, and was consistently rewarded by her clear understanding of what I was trying to say and her unfailing perception of how to say it better.

Ongoing conversations with Dale Tomich about parallels and differences in his work and mine on materially and spatially informed understandings of world-system history have been a delight and an inspiration for over a decade; they have shaped this work directly and indirectly. He as well has understood, often more clearly than I have, what I was up to and what parts I needed to bring into close focus. Henry Tom's insistence that I cut extensive parts of the book was painful, put me through enormous effort, vastly improved the book, and has left me with intact chapters that I hope will become the deeper analysis of Carajás that the Amazon and its population so deserve.

My wife, Dena Wortzel, in addition to listening to my endless and often changing talk about the book and reading and editing large sections of it, has worked lovingly and untiringly to keep me alive and cheerful while I was writing it.

Different parts of the data that inform this analysis were gathered on grants from the U.S. National Science Foundation, Brazil's Conselho Nacional de Pesquisa, the Howard Heinz Endowment, the University of Wisconsin Graduate School, the University of Wisconsin Global Studies Institute, the World Wildlife Fund, and the Sloan Foundation. I am grateful for the kind support of all of these institutions.

—Stephen G. Bunker

Globalization and the Race for Resources

CHAPTER ONE

Matter, Space, Time, and Globalization

An Introduction

Even though our story encompasses centuries, it begins a mere two decades ago, in a sparsely populated outpost deep in the Amazon rain forest—one of the largest, most complex, least developed, and least populated ecosystems in the world. There, in a project so huge that World Bank funding required the coordination of loans from six major industrial nations, Brazil's Companhia do Vale do Rio Doce (CVRD), the world's largest iron ore exporter, inaugurated Carajás, the world's largest iron mine, by sending its first shipment of ore on the world's longest mine-dedicated railroad to the world's largest iron-shipping port. CVRD had built both the railroad and the port for the sole purpose of exporting Carajás's ore. At the port, huge dump loaders tipped the ore out of the railroad cars, and a system of cranes and conveyor belts loaded it into one of the world's largest ships. That ship carried it halfway around the world, to Japan, where it would serve as raw material for some of the world's largest steel mills.

The mine's size, its distance from the world's established industrial centers, and the scale of its transport system—vehicles, vessels, roads, rail lines, port, docks, and machinery for loading and unloading are all enormous, powerful, and very heavy—make the inauguration of the Carajás mine one of the largest and most significant in the thousands of separate capital-intensive and scale-expansive steps that have moved the world economy toward globalization. Carajás quintessentially blends the technical scale and the spatial scope that characterize global projects. All of its systems, from mine to port to ships, were built in gargantuan proportions in order to reduce the per-ton cost enough to make its iron ore competitive in international markets.

A mine as big and as remote as Carajás is possible only because of the multiple technological, organizational, political, and financial innovations that paved the road to globalization. In that sense, the mine emerged from a long evolution of interconnected tools, skills, and institutions devised to achieve and sustain economies of scale. Carajás's history is about the local consequences of this globalizing evolution, but, as the most recent pinnacle of historically reiterated and ever-grander cases, it summarizes the historical tendencies—ecological, social, and financial—of extractive projects worldwide.

Iron's very low value-to-volume ratios in contemporary metals markets lend the global scale of the Carajás mine a particularly dramatic significance. Since the Industrial Revolution, iron and coal have constituted the largest volume of inputs into industrial manufacturing. Of these inputs, they have the lowest cost per pound. Until the 1950s, low value-to-volume ratios for iron and coal limited the markets for these voluminously consumed raw materials to the regions immediately surrounding the steel mills themselves. Even so, for the steel mills transport frequently represented as much as 60 percent of the cost of iron ore.

By the 1960s, the booming Japanese steel industry had broken these limits. Japanese steel mills were importing huge volumes of iron and coal from Australia, over 5,000 miles away. No major steel company had ever transported its main raw materials such distances. Reducing transport costs from Australia enough to make Japanese steel competitive required revolutionary innovations and economies of scale in the power, strength, and size of ships, loaders, and ports. Such economies of scale in transport could function only if mines of unprecedented size were dug into huge deposits of high-quality ore to fill ships—on a regular basis and with minimum delay in harbor—three to four times bigger than the largest dry-bulk carriers in existence only twenty years earlier. Such deposits are rare and widely scattered. By this time those closest to industrial centers had already been mined and depleted.

In the 1980s, less than thirty years after Japan's bid to dominate the world steel industry had expanded iron and coal markets and transport technologies from continental to hemispheric scale, CVRD—with logistical, political, and financial support from the Brazilian government—negotiated global financing for a vastly larger mine and a transport system in Carajás. CVRD achieved such economies of scale there that its iron ore could compete in global markets 5,000 miles away, in northern Europe, and over 12,000 miles away, in Japan and South Korea. To haul the iron ore to the ships that would make these long voyages, CVRD constructed a heavy railroad across 890 kilometers of difficult tropical terrain, cut by many rivers and subject to heavy rains. This railroad cost far more to build than any of the shorter, easier lines—none of which crossed more than 400 kilometers of gently sloping dry ground—that connect the various Australian mines to their ports. The same spatial and topographical barriers increased the cost of constructing the mine as well. Only the huge size and the high quality of Carajás's deposits could justify the enormous investments required to extract and transport iron from such an inaccessible location.

Carajás is a prime example of the centuries-long and still accelerating trend of world industry to consume ever-greater amounts of raw materials from ever-larger deposits, ever-more distant from the industrial centers that transform them into commodities. For at least half a millennium, multiple individual firms, and by extension the national economies of which they are a part, have competed for market share by instituting progressively greater economies of scale. Economies of scale lower the unit cost of production by producing more units, which generally requires expanded investment in larger, stronger machines so that the total cost is increased to achieve a lower cost per unit produced. Such economies of scale, though, create diseconomies of space. Expanded production transforms greater volumes of more different types of raw material. Increasing the amounts and types of matter transformed requires procuring and transporting it across greater distances.

The causes of this contradiction lie in simple but inexorable physical laws and in more contingent, but very powerful, economic forces. Humans cannot create matter, so they must extract it from natural sources. Raw materials are produced at naturally fixed rates. Most minerals are produced very slowly in specific places and are stored there in finite amounts. These sources are scattered around the globe. Firms tend to exploit the most proximate or convenient sources first, so they have to go to more distant, and therefore more costly, sources as they deplete the nearby ones, or when they devise technologies that require higher grades or larger deposits of ore. Ongoing competition leads to reiterated expansions in the scale of production, so progressively greater volumes must be transported from sources that are larger and more widely distributed and thus on average more distant and more costly.

Solutions to this contradiction also conform to physical law and to economic rules. Firms, sectors, and states strive to overcome the diseconomies of space by building larger, stronger, and more efficient vessels and vehicles. Ports, rails, bridges, loaders, and other infrastructural elements must then also become larger and stronger to handle the greater weight and size. The geometry that relates volume to the cube of the radius, but surface (or hull, in the case of a boat) only to its square means that inertia, momentum, resistance, or drag all increase more slowly than volume transported. Building a bigger ship therefore increases cargo capacity much more than it increases the amounts of fuel and labor needed to operate it. The result of these geometric differences—i.e., that bigger boats carry more cargo more cheaply—stimulated historically reiterated cycles of technological innovation that made it possible to build and navigate progressively larger vessels. In each cycle, production economies of scale created material diseconomies of space that in turn generated transport economies of scale that, by supplying more and better raw materials more cheaply to industry, enabled further economies of scale in production.

This process is self-reinforcing and reiterative. Building larger transport vehicles and expanding the required infrastructure—docks, ports, railroads, highways, derricks, cranes, and conveyor belts—transform greater volumes of matter, consume greater amounts of energy, and employ more labor in a transport process that provides more and cheaper raw materials for production. Lower production costs stimulate higher consumption rates, expanding production, wages, and markets even further. Interdependent economies of scale in transport and production thus directly increase demand for material and machines, and additional wages increase demand for commodities. Increased demand enables competitive firms to expand their production economies of scale even further. New and even greater economies of production scale then create new and even greater diseconomies of space, thus stimulating further increases in transport scale, which set off new and ever-greater expansive cycles.

The endeavors of competing firms and nations to procure cheap and stable sources of raw material drove the historic reiteration of these progressively larger cycles, which eventually created the huge scale of contemporary extraction and transport systems. In this book, we consider how abstractly general physical processes and relations intersected with place-specific social and economic history to generate these scale-enhancing solutions. We examine the design, financing, construction, and operation of the mine at Carajás as the most recent peak of the progressive increases in scale toward which raw material extraction has tended for over six hundred years. In order to understand the history that led to the mine, we compare its scale, its operation, and its trade relations with the scale, operation, and trade relations of other Amazonian extractive economies over the last five centuries. Then we take what we have learned about the sequence of extractive economies in the Amazon to frame a comparative historical analysis of the five nations—Portugal, Holland, Britain, the United States, and Japan—that successively rose to dominate world trade, as well as that of the Amazon, over the past six centuries.

As we compare different extractive economies over time, we see that social, technological, material, and spatial processes, operating at very different levels of abstraction and across very different domains—from the general and global to the specific and local—intersect differently to configure specific ecosystems and the extractive economy organized to appropriate particular material forms from each. This intersection enables us to use what we learn about the history, sequence, and internal dynamics of extractive economies—both in the Amazon and elsewhere—as a prism to examine the material, spatial, and social features of the national economies that rose to dominate world trade.

We then use this prism to frame a comparative history of the trade-dominant nations' raw material access strategies. We find that trade dominance depends on combining technologies that enable competitive economies of scale in production with others that enable cheap access to and transport of raw materials in the volumes that the new economies of scale require. Each nation that dominated world trade managed to expand and adapt existing technologies—and in some cases to invent new ones—that integrated general physical and material processes with its place-specific social organization effectively enough to achieve favored access to raw material. Each of them remained dominant until another nation integrated social and natural processes in ways that generated even more powerful technologies, financial and social institutions, and political systems.

As our comparative analysis moves from extraction to industrial production, from local to global perspectives, and from physical to social process, we find general patterns in the ways firms and states innovate and strengthen technology, finance, and politics in order to secure cheap and stable access to the raw materials they use most. Within these patterns, though, we note that the social, financial, and technical organization of extraction and transport must conform to the material and physical attributes of each material and of the place from which it is extracted.

These nations became dominant by successfully resolving the tension between economies of scale and the costs of space. Each of them did this by devising new technologies that enabled it to build and operate larger vessels and vehicles across spaces that were previously inaccessible to high-volume, low-cost transport. Such economies of scale reduced the costs of transport but were only viable if the scale of extraction and the scale of finance rose with the scale of transport. Over time, the accumulation of these reiterated resolutions combined (a) intensification of social production, (b) extension of integrated material and financial processes and relations across space, and (c) physical expansion in the scale of extraction, transport, and production. These expansive dynamics drive globalization.

Thus, globalization emerges as the intensification and expansion of material processes of production and exchange require greater volumes of a greater variety of materials, which may be available only by extending extraction and transport across ever-broader portions of global space. This material is produced in physical regularities and processes that are governed by nature and configure differently in different places. Globalization results from the ways that matter and space, as unchanging components of nature that configure differently in site- and time-specific forms, interact with economics and politics as locally and temporally specific activities of society.

Technology, socially devised to enhance the utility of human interactions with matter and space, mediates between society and nature. The groups that control technology tend to use it in ways that enhance their social power over matter and space—and thus over other groups and the environments in which they live. Matter and space are bound by immutable natural laws of physics and chemistry and biology. Technology is socially produced in ways that cumulatively increase its capacity and scope over time, but each innovation works only if it conforms to the immutable laws that govern the matter and space with which it is designed to work. All material processes—whether naturally or technologically initiated—necessarily occur in some local place. Understanding the anomaly of Carajás—an enormous and very local hole in the ground, in a pristine jungle remote from industry, dug by enormous machines to extract huge volumes of raw material to send to industrial centers halfway around the world in huge trains and ships—will help to explain how local processes are materially intensified and spatially expanded toward globalization.

How Local Process Drives Global Systems

The extraction of natural resources for globalized raw material markets imposes significant disruptions and inequalities on local ecological and social systems. Larger-scale extraction tends to increase the damage to local systems. Understanding these dynamics requires close attention to the complex inter-weavings of matter and space with economy and society.

This is no easy task. In the advanced postindustrial economies of Europe and North America, most of our lived experience of economic processes con-sists primarily of consuming the products of increasingly homogeneous, in-creasingly large-scale industry. Most of us live and work in globalized econo-mies in which process and product are increasingly distant from the sources of the raw materials they transform. Few of us directly experience the processes whereby multiple naturally produced heterogeneous material forms, extracted and transported from many different places, are combined to produce the standardized commodities we purchase and use.

Marx explained that mainstream theories of capitalist economies fetishize commodities so as to mask the social relations that produce them. More recently, Willis (1991) pointed out that the growing distance between produc-tion and consumption that occurs as firms with markets in wealthy countries move production to poorer countries makes it even more difficult to perceive the labor in the commodities we consume.

Ideology and distance make it difficult to discern the social inequities and exploitation in commodities. Perceiving and thinking about how extraction of the raw materials in these commodities disrupts natural production is even more challenging. By the time we experience industrial products as com-modities, the raw materials they contain have been transformed and combined in ways that make their natural forms and shapes, as well as their place of origin, extremely difficult to identify.

Incorporating local space, matter, and society into our concepts of the global in analytically compatible ways therefore poses a major challenge for contemporary scholars of both world-systems and globalization (e.g., Chase-Dunn et al. 2000; Tomich 2004; Robinson 2001). Most of these authors ignore both the materiality and the locality of production. They assume the global as their point of departure and attempt to incorporate the local into it.

We aim to reverse that logic. We take into account and theorize the interaction of natural and social processes. In other words, within the economic logic of global markets we integrate the "eco-logic" or materio-spatial logic with the "socio-logic."

Matter, Space, and the Logic of Production

All production is profoundly material and local. Understanding the expansion and intensification of the social, material, and spatial relations of capitalism that have created and that sustain the growth of the world economy requires analysis of how material processes of natural and social production interact with each other and with space. Such analysis necessarily encompasses the ways that technological, social-organizational, and political innovations change these interactions over historical time. This means that we have to examine how multiple heterogeneous and spatially distinct natural systems produce the diversity of material forms that constitute an increasingly homogeneous global economy. In order to take account of space as materially differentiated and matter as spatially differentiated, we must incorporate topography, geology, hydrology, and climate, as well as absolute and geographic distances between places—which themselves result from the intersection of these forces—into our analysis of how trade-dominant nations assure cheap and stable access to the volumes and types of materials they need.

Many new technologies have used new and different combinations of raw materials, so this examination will require us to consider historical changes in the ways that technology and social organization respond to and mediate between matter, space, and human purpose. In other words, we will look at how historic changes of technology emerge from social action but also how they must be shaped to and mediate between the abstract laws of physics and chemistry and the more concrete regular processes of geological, seasonal, hydrological, and biological cycles and forms. Technological change, then, remakes society's relation to nature. In the process, the accumulating power of these technologies strengthens the societies and economies that generate them while leaving the societies and economies that provide their raw materials weaker and more dependent.

As capitalist economies of scale expand, development and implementation of the technologies that make them work become increasingly costly. As the scale of these technologies increases, the circuits of capital that initiate them expand and accelerate. Generation of and control over finance expands the relative power of the societies that generate new technology. This means that finance and technology drive each other toward globalization. We therefore need to construct a preliminary model of how technology and finance respond to material and spatial constraints as these affect capitalist competition for trade dominance.

A Preliminary Model of Globalization

Globalization—though identified as a recent or novel phenomenon by Sklair (2000) and others—in fact results from cumulative materio-spatial expansions and intensifications, which are driven historically by reiterated economies of scale made possible by technology. In nations competing for world trade dominance, firms and sectors often collaborate technically, financially, and politically with other domestic firms and sectors and with the state to develop and implement these technologies—even as they compete with each other to capture greater market share and higher profits. This collaboration generates dramatic episodic increases in economies of scale in industrial production, raw material extraction, and transport.

Economies of scale expand, intensify, and diversify the consumption of materials needed to produce greater quantities of a larger variety of commodities. Constructing larger, faster, stronger machines and infrastructure for the technologies that make these economies of scale possible also increases and diversifies the consumption of greater volumes of a greater range of raw materials. Each introduction of new technologies that used new material inputs—first iron and coal, then steel, then a series of other materials such as manganese and petroleum—quickened both the expansion of volumes and the diversification of types of material consumed by industry.

Volume, diversification, and precise specification of the types of material that scale-enhancing technologies require accelerated particularly dramatically as thermally powered engines—progressing from coal and water for steam to the internal combustion of petroleum to nuclear reactions of uranium—replaced animals, wind, and water as the prime motor force in industry. All of these engines produce and must contain pressure and heat; the more powerful the engine, the greater the heat and pressure. The greater the heat and pressure, the stronger the material needed to contain and resist them. To make stronger materials, more metals must be precisely processed and alloyed.

These material requirements intensified as larger, stronger, faster machines expanded technologically achieved economies of scale and speed. Humans discovered new ways of combining and processing the physical and chemical properties of different raw materials in order to achieve the additional strength, flexibility, and resistance needed to create the larger, stronger, faster machines, vessels, and vehicles that make the expanded economies of scale possible.

Resistance to intense heat and to heavy pressures requires alloys of specialty metals available in few places, so the material requirements of the new technologies extended searches for sources to more distant sites. Expansion, intensification, diversification, and more precise specification combined to make new technologies increasingly dependent on more and larger uniformly higher-grade deposits of raw materials across broader spaces.

This is not a new phenomenon, though its effects have accelerated dramatically over the past 150 years. Trade relies on transport; the complex interplay of matter, space, and technology is as old as the competition to dominate trade. Expanding the size of sailing ships to cheapen freight costs in the seventeenth century required larger, stronger, longer, appropriately curved oak boards for the keel; longer, stronger straight oak for planking; and larger, stronger, lighter pine for masts able to sustain the additional wind pressure on their much broader sails. Large, high-quality pines provided reach and strength without adding the extra weight above the water line that would make the ship more prone to capsize. To gain speed and maneuverability without sacrificing strength and cargo space, Dutch boat builders learned to form the hull from a single sheet of oak planks precisely cut, then joined and caulked around a skeleton frame of much thicker timbers. This technology required a wide variety of precisely specified shapes and sizes of different types of wood, together with particular kinds of tar for caulking and long, strong fibers of flax for making the sails. All of these materials grew in different parts of the world, so the new technologies that supported Dutch competition in world trade required the complex coordination of differentiated matter over broad space. British shipwrights later adapted the innovations that had enabled the Dutch to dominate trade by building cargo ships that were more efficient and easy to manage. The British, however, also built strong, maneuverable warships to overcome the advantages that carrying efficiency provided to Dutch merchants.

Similarly, the Japanese engineers who designed and built the supertankers and the enormous dry-bulk carriers that enabled Japan to challenge the U.S. domination of world trade at the end of the twentieth century relied on large, cheap supplies of steel of much higher tensile strength and flexion than could have been mass-produced fifty years earlier. Producing large volumes of uniformly high-grade steel cheaply enough to enable Japanese shipyards to capture over half of the world market required access to huge deposits of iron and coal of precisely specified purity, grade, carbon and sulfur content, hardness, and moisture content. As with wood, flax, and tar, these material features of iron and coal, as well as the other mineral inputs for high-tensile steel, were naturally produced in different parts of the world.

Engines to move these huge ships require equally high grades of matter. Similarly, the engines and bodies of supersonic jet–powered aircraft must resist far more heat and stress than the DC-3s of the middle of the twentieth century were designed to withstand. Cobalt and titanium, necessary for supersonic engines, are extremely rare in nature. Capital has had to seek them in remote and difficult places. The dependence of national states on the stronger tax base and military power—impossible without these raw materials—makes them ready and anxious to assist financially, diplomatically, and militarily in this search.

Matter and space are naturally given. Technologies that mediate between them and human actions and goals are socially created, but they can only achieve the human goals for which society invents and finances them if they conform to the natural (i.e., biological, geological, locational, physical, and chemical) features of the raw materials they transform. As human economies expand their political domains from communities to nations to continental unions, competition between these units to dominate trade—first locally, then regionally, nationally, and finally globally—drives searches for technologies of production and transport that enable increased economies of scale. The technologies that make these increases in scale possible only work with particular material forms that are naturally produced in the appropriate quality and quantity in particular places.

These different features of raw materials are naturally produced in specific locations. They form the material bases and the spatial context for the financial and political competitions and collaborations that have driven globalization, not just over the past two centuries of competition to improve iron- and steel-based technology, but through the earlier centuries when competition to dominate world trade required scale-enhancing technological innovations in using wood, sail, and wind for production and transport.

To summarize, in production as in transport, and with relatively few exceptions as capitalist economies have expanded, both the larger, stronger machines and the commodities they mass-produce require higher-grade raw materials with more precisely specified chemical and physical properties. As the volumes produced and sold expand, so do the volume and value of capital invested in building plant and machinery and maintaining them. As the scale and speed of processing and production increase, adjusting larger machines for variations in purity, grade, and physical composition of raw materials becomes increasingly costly—both in the technical operation itself and in the production lost during the time any given machine is down. Capital therefore seeks large and uniformly high-grade deposits of many different materials. The larger and purer the deposits, the more broadly distributed they are in space. Extractive economies thus become more dispersed, while productive economies become more agglomerated.

This means that the natural logic of matter as it is produced in space creates a contradiction with the social logic of capitalist economies of scale. The contradiction is resolved and then reiterated on a larger scale with each cycle of technological innovation in expanded factories and larger machines. These material and technical dynamics combine to assure that expanded consumption of matter requires transport of ever-greater volumes from the diverse locations where different material forms are produced. The end result, reiterated in each struggle for trade dominance, is that material intensification and the economies of scale that drive it generate the greater expense of transporting larger cargos across greater space.

Each such iteration surpasses in scale and scope the material and spatial problem and the technological solutions of the previous cycle. Each cyclical reiteration of these processes (1) expands the space in which national firms procure raw materials, which thereby (2) stimulates development of stronger, faster, more capacious technologies—ships, trains, and loading equipment— and larger, more extensive infrastructure—ports, rail lines, and warehouses— for their handling and transport, which themselves (3) reduce the cost of exporting products and so expand the space in which each national economy can competitively market its products and thus (4) advance the globalization of the world economy while expanding its reproduction of capital.

Globalization can therefore best be understood by comparative historical analysis of the discrete, sequentially expanding cycles of contradiction and resolution between economies of scale and diseconomies of space. In each cycle, the technological innovations that create, and then the innovations that resolve, this contradiction must be adapted to the local configurations of the matter they incorporate and transform, to the space from which the matter is extracted, and to the space between the location of extraction and the location of transformation. The interaction between the socially designed technologies used and the naturally produced matter and space on which they are used shape the opportunities and the constraints that mold local systems of extraction, transport, and production even as these feed into global systems of trade and finance. (Even transport, the movement of matter through space, is at each moment as local as the location of the vehicle and as the provenance of the capital sunk in the built environment at each point in the space traversed.) These global systems are themselves shaped by nations competing to dominate trade. In order to analyze these complex interactions as they intensify and expand historically, in the next section we combine the most coherent theoretical treatments of space, matter, and technology during successive systemic cycles of accumulation in ways that enable us to reinterpret the standard treatments of trade and hegemony. Then, in the sections that follow, we extend the materio-spatial paradigm we derive from this combination to a framework that enables a systematic approach to the intersection of physical law, material process, social structure, and human agency. This framework allows us to move beyond the explanatory primacy attributed to the social in even the most coherent treatments of space, matter, and technology.

Insights from Other Analyses of Matter, Space, Time, and Capital

Wittfogel's (1929) compilations of Marx's analysis of natural systems in the production of use values, Marx's theory of rent (Bunker 1986, 1992), and Merchant's (1983) discussion of Francis Bacon's and Agricola's views on the imperative that all human uses of natural systems succeed only as their technology responds to material and chemical attributes of these systems—these explanations all point to the ways that the physical and chemical attributes of different kinds of matter enter into all human technological improvements. The very diversity of matter dispersed across space provides the material and spatial means of the secular intensification, proliferation, technification, and expansion of both the types and volume of material production that drive the system's dynamic of growth and change. These considerations show us ways to move beyond Wallerstein's (1982) tripartite categorization of the world-system and to correct his tendency to use a categorical logic instead of materially and spatially grounded mechanisms to explain the dynamics of growth and change. Doing this, in turn, will enable us to achieve an adequate specification of particular or local economies.

Marx's rent theory specified the social and political implications of his abstract affirmation that nature and labor are inextricably interdependent in the production of commodities. Marx (1894) showed that the quality and price of raw materials determined the productivity of labor and the organic composition of capital. He then showed how, under capitalism, rents tend to be differentiated so as to correspond to the natural fertility of the ground for which they are paid. The relative fertility of soils in different places determines the contribution of the natural products of that ground to the productivity of labor and thus to the rates of profit on the goods cultivated on or extracted from it. Differential rent is thus set according to the contribution of particular lots of ground to the productivity of the labor employed on that ground. Marx used agriculture to work out the logic of rent, but he assured us that the concept of fertility can be applied to timber forests or to mineral deposits as well. However, the vastly more expansible economies of scale in mineral-based technologies importantly change the costs, the profits, the scale, and therefore the rents for mining economies quite differently than they do for agriculture.

Harvey (1983) partially clarifies these differences without realizing or theorizing the underlying dynamics. He extends his own concepts of the built environment combined with a synthesis of Von Thunen's (1966) central place theory into an appreciation of Marx's theory of rent. In this effort, he explicates and updates Marx's insights into capital's dependence on the state to redistribute the huge sunk costs of transport technologies and infrastructure to include the various economic and social agents that benefit from them. Harvey's concern with urban processes limits his focus almost exclusively to labor and capital. Marx's interest in and concern with agriculture, forestry, and mining helped him incorporate the full trinity of land, labor, and capital into his analysis of rent, with land representing the contribution of natural forces of production to the creation of value (see also Coronil 1997: 45–48). We must extend Harvey's treatment of the built environment to incorporate fully this trinity while also including Harvey's useful updating of the Marxian insights about the state's need to collaborate with capital to overcome the linked problems entailed in building transport systems, which must (1) serve multiple competitive interests, all of whom will benefit and none of whom wish to assume the costs of building the necessary infrastructure; (2) impinge on multiple properties in land, all of whom have to allow passage over their land if the system is to operate; and (3) be financed by investments that exceed the capacity of single capitals.

Marx very clearly emphasized material and spatial dynamics in his assessment that the costs of capital sunk in infrastructure had hugely expanded as coal, iron, and steam made new means of transport possible. The materio-spatial, technical, and financial dynamics of iron, coal, and steam provided use values that could only be realized as new transport technologies moved them cheaply in bulk from their dispersed natural locations to the social locations of the increasingly agglomerated urban-industrial centers that supported the expanding scale and volume of machino-facture. Because Harvey's explanation privileges the internal dialectic of capital as autonomously causal and fundamentally financial, however, he does not fully appreciate how much of capital's internal dialectic emerges from its material history, or what we might call the external dialectic of capital with nature. The environment socially built to enhance the productivity of labor (and the returns to capital) does so by facilitating and cheapening capital's access to nature in the form of raw materials. The cheap bulk transport that the built environment makes possible may lower ground rents paid for this access by vastly expanding the different sources of raw materials economically available to capital. Harvey misses both of these materio-spatial implications of Marx's rent theory, but we will see in the next paragraph that his theory is open to further specification of materio-spatial processes. Once that is done, we can incorporate it into world-systems theory in ways that enrich and unify all three theories as components of a theory of globalization.

Harvey's insights into transport and the built environment usefully complement Innis's (1933, 1956) conceptually different but analytically compatible explanations of the costs, dynamics, and politico-organizational consequences of transport systems devised to satisfy core demands for peripherally extracted raw materials. Reading Harvey and Innis together clarifies how transport systems function as increasingly capital-intensive, debt-creating, state-forming instruments to articulate dispersed site-specific raw material sources with concentrated centers of industrial production, capital accumulation, and political power. Materio-spatial analysis of the financing, construction, and use of specific transport systems' technology and infrastructure in specifiable times and places reveals the specific interests and activities of different groups in both core and periphery. It can explain how transport technology and infrastructure progressively cheapened and accelerated the consumption of natural resources in each cycle of world-systemic material intensification and spatial expansion.

Materio-spatial analysis of (1) specific extractive economies and political formations of peripheral zones, (2) the technologies, composition of capital, organization of labor, and state formation and reformation in core economies, and (3) transport and communication systems between them reveal how these processes have driven progressive globalization. These attributes, along with the secular evolution of the technological and political systems they generate, mold the strategies, costs, spatial extents, and profits of core initiatives to exploit multiple diverse peripheries for their varied natural resources. The transtemporal continuity of interactions between social, spatial, and material systems explains the sequentially cumulative processes that have expanded and intensified productive and commercial relations through each systemic cycle of accumulation.

This continuity belies any claims, such as those already noted by Sklair, that globalization is a novel or recent phenomenon. Rather, it is the latest and perhaps the ultimate stage in the materio-spatial processes that have evolved and accumulated sequentially through diverse activities and interactions of identifiable groups and organizations operating in particular locations and territories (Bunker 1996; Bunker and Ciccantell 1995a, b, 2003a, b; Ciccantell and Bunker 1997, 1999). Paradoxically, it is only called globalization as these processes reach their global limits.

This continuity also belies the "destatization thesis," whose proponents claim that globalization supersedes the nation and the nation-state and makes them both irrelevant and misleading as units of observation or analysis. The role of the nation-state, its relations to its internal politico-administrative units (such as federated local states, counties, or municipalities) and to other nation-states, its relative autonomy from and collaboration with dominant classes and their financial and business institutions, and its treatment of the working classes, in addition to varying between nations at any moment in time, have changed over time with the technological, economic, financial, and political changes that have moved the world-system through successive cycles of systemic accumulation that led to globalization.

That the most recent stages of globalization diminish the capacity and autonomy of nation-states in the periphery does not mean that we can dismiss the state as an important actor. Earlier theories of internationally uneven development considered the interactions between dominant political econo-mies and the colonial or dependent states whose power and autonomy they restricted as an illuminating object of analysis. Similarly, the case of Carajás shows that the relations of national and local states—to each other, to national and international firms, and to other nation-states—critically mold the ways that local economies are incorporated into global markets. A close analysis of how states respond to or are manipulated by different agents acting in a variety of power domains provides critical insights into the interaction of local and global processes.

Raw Materials, Trade Dominance, and the New Historical Materialism

Patterns of world trade in raw materials change dramatically during the rapid rise of a new competitor. Holland and England each transformed, and then struggled over, trade in timber to build ships at the beginning of their ascent; Britain, Germany, and France struggled similarly over iron and coal as those materials became the most voluminously used in world industry. The United States had enough timber, and later enough iron and coal, situated conveniently enough to multiple waterways to enable it to become the world's dominant producer, first of wooden, and then of steel, ships. Japan revolu-tionized global ocean transport in order to create and control markets for iron and coal.

Each competitor nation strove to reform these trades to their own loca-tional advantage. The Dutch devised ships, banks, and markets that max-imized the advantage of their location between the bulky raw materials they procured in the North and Baltic Seas and the wealthy Mediterranean markets where they sold them. The British elaborated complex systems of colonial control and military might to secure access to the resources of the Atlantic and the Indian Oceans. The Americans extended and cheapened overland trans-port to connect the Atlantic and the Pacific Oceans and went on to dominate global trade. Japanese firms and state agencies continued the pattern of mate-rial intensification across broader spaces. Like their predecessors, their rapid industrialization and ascent to preeminence in the global economy fomented global shifts in raw material extraction, processing, and trade. Under 20 percent of all iron ore mined in 1960 entered transmaritime trade; under Japanese dominance two decades later, over 35 percent was shipped overseas for processing. Ton-miles of iron transported increased over 600 percent during the same period. Coal was also transformed from a locally and regionally traded commodity to a globally traded one; ton-miles transported increased 1275 percent between 1960 and 1990.

Two central propositions emerge from our comparison of the five nations that rose to dominate trade: (a) securing reliable access to a variety of cheap raw materials is the most complex challenge that any rising national economy confronts (Bunker and O'Hearn 1992; Bunker 1994a; Bunker and Ciccantell 1994; Ciccantell and Bunker 2002); and (b) this challenge becomes more complex over time, as the scale, scope, and throughput required for effective competition drive the long historical trajectory to globalization. Solutions to this challenge require technological, financial, organizational, and institutional innovations capable of restructuring relations at previously unprecedented degrees of close-coupled complexity, scale, and scope within and across firms, sectors, and the state. These newly structured relations must be compatible and dynamic both domestically and internationally. They must also be flexible enough to respond effectively to the responses of other core nations and extractive peripheries to changes that the rising economy's access strategies and developmental trajectory stimulate in the world economic system.

Taken together, these propositions illuminate how the material intensification and spatial expansion that result from each successful national ascent to trade dominance accumulate sequentially through successive systemic cycles of accumulation. These key dynamics simultaneously advance globalization and exacerbate the differentials in state, corporate, and financial power, skill, and organizational complexity that underlie the growing inequalities between national economies.

Any rising economy must adapt its solutions to a larger, politically and technologically more complex world economy than had confronted earlier ascendant nations, but the parameters of effective raw materials access strategies persist across all cases. In any rising economy, these strategies must respond to and take advantage of contemporary technological, geopolitical, environmental, and market conditions in the rest of the world and of the nation's position and location within that particular global economy. They must also adapt these technologies and their own social and financial organizations to coordinate the physical characteristics and location in space and in the topography of the various raw material resources actually or potentially available with the physical characteristics and location in space and topography of the national territory.

The solutions to the raw materials problem require the coordination of multiple physical and social processes across geopolitical and physical space with domestic relations between firms, sectors, the state, labor, and new technologies. The solutions to these problems must also accompany or even precede industrial competitiveness. Therefore, these solutions require and stimulate complex processes of learning and of institutional change that fundamentally mold the organization of the national economy at the same time that they change international markets and the rules binding participants in them.

We extend to a global stage Alfred Chandler's (1977) demonstrations of how the railroad and the steel industries created templates and dynamic forces that transformed the shape of U.S. corporations. Chandler, though, used a historical methodology, which limited the nomothetic potential of his insights. We propose, in contrast, to interpret the economic, social, and geopolitical history of raw materials access strategies and of the transport innovations and investments needed to implement them through first principles of chemistry and physics. Chandler's insight that heavy materials and the systems devised to transport them rapidly over long distances formed the crucible of the U.S. economy can be abstracted upward to reflect not merely a national particularity but rather a set of physical determinancies that structure the ways that procurement and transformation of matter expand in volume and in distance transported as the industrial economy itself expands.

Our approach, which we call the New Historical Materialism, allows us to use first principles of physics, chemistry, biology, and geology to explain why the solutions to the problems posed by procuring, extracting, processing, and transporting the heaviest and most voluminous of raw materials has generated social attributes essential to each economy's rise to global dominance. By first framing our historical analysis within the regularities governing material, temporal, and spatial characteristics and processes and then showing how these physico-spatio-temporal processes intersect with and shape economic regularities and mechanisms, we can extend the lessons of particular cases to more abstract principles. This parallels North's (1958, 1968) approach in the New Economic History, but there is far more consensual validation for the table of elements or the laws of physics more generally than there is for the first principles of classical economics or for sociological definitions of the state or the firm. Indeed, historical materialism is fundamental to explanations of combined and uneven development and thus is more appropriate to the relative and relational analysis implied by world-systems approaches.

In devising this approach, we have updated Marx's use of then contemporary understandings of biology and organic chemistry to show how socially devised technologies invalidated the formulas that Malthus used to project lower rates of increase for agricultural production than for human populations. We also follow—indeed, update—Marx's understandings of the mechanical rules and physical laws that governed the amount of pressure per square inch that could be safely contained in a steam-engine boiler. Marx based his theory of differential rent on notions of natural soil fertility and of how it could be technically enhanced. He extended his theories of differential fertility to apply to ore quality in mineral extraction.* We now have access to far more precise formulations of physical, chemical, and biological laws than existed when Marx first advocated material history as a means of demystifying political economy and invoked vaguely specified notions of organic metabolism, but Marx was clearly searching for ways to integrate material and social process in his analysis.

Framing the question of how economies ascend in the world-system within a rigorously materialist logic validates a focus on how the cost and quantity of raw materials form the basis for economic and social development. The physical and chemical characteristics of matter, energy, space, and time impinge far more directly, and far less flexibly, on raw materials economies than they do on sectors farther downstream (Bunker 1992). The total volume of raw materials is necessarily greater than the volume of the products made from them. The diversity of their molecular structures and the heterogeneity of their chemical compositions are also necessarily greater than those of the industrial inputs derived from them. The distances and the topographic and social organizational differences between the sources of raw materials are greater than those between the factories that produce these inputs. Greater volume, greater distance, greater complexity, and more heterogeneity combine to create the

---

*Much of the first third of Volume III of *Capital* is devoted to Marx's accounts of the practical experiments of and the dialogue between factory engineers and public safety inspectors struggling over the opposed imperatives of increasing engine power by increasing steam pressure and of protecting the safety of workers and plant. The middle third of Volume III addresses the ways that soil fertility, natural and socially manipulated, differentially affects rates of profit. Though only roughly edited and integrated, both of these sections of Volume III deal with issues of the effects of naturally produced features on political economy.

broadest scale and scope for the most fundamental of economic processes—the increase of profit and market share through technical and organizational innovations that reduce unit costs of raw materials.

The solutions to problems of bulk, throughput, physical complexity, and cost in raw materials procurement and processing require coordination of domestic and international processes. These must be compatible both within and across particular raw materials. Raw materials industries thus tend to generate the domestic reorganization of corporations, sectors, and the state as well as the extraterritorial understandings, rules, and practices that underlie the operations of a world-system.

The most voluminously used raw materials are those that bear weight, resist force or heat, and provide energy, and so are most extensively incorporated into infrastructure, machinery, and transport. Infrastructure, machinery, and transport facilities must all increase rapidly during the initial stages of industrial development. The bulk and weight of wood, and later, of iron, coal, and oil are vastly greater than those of the other materials that are procured and transformed. The synergies between their procurement, transport, and transformation at the accelerating levels of throughput, scale, and scope that a rising economy must achieve create the challenges and conditions for the social, economic, technological, political, legal, and educational innovations required for the ascent of a national economy within the global economy.

This was true of wood and ships in Amsterdam; of iron, coal, rail, and steamships in Great Britain; and of iron, coal, oil, rail, ships, automobiles, and trucks in the United States and later in Japan. Particular technological, geographical, and relational solutions varied because of the huge changes in scale, scope, technology, and geopolitics during the very different periods of these different national ascents. Technological solutions in both the transport and the transformation of matter and energy conform to specifiable and unchanging chemical and physical laws and regularities. Space and topography act as geographic obstacles to transport whose effects can be precisely measured for each transport technology. Technology and geography thus provide baselines within which regularities and mechanisms with fairly high explanatory status can provide a context for the examination of the different relational solutions that are viable at different moments of competition for national ascent within the world economy. Our challenge is to understand how the regularities of technology, space, topography, and matter intersect with the contingencies of social process and organization.

Universal Laws, General Rules, Regular Patterns, Historical Contingency, and Chance

We do this by examining the ways that society and nature interact through the social creation and use of technologies that can only perform their intended functions if they simultaneously take advantage of and conform to the diversity, regularities, and limitations of naturally determined material processes. Without the diversity of material forms available in nature and the regularities of material process, humans could not have evolved as they have, let alone devised the complex and powerful technologies that serve them. At the same time that it makes these technologies possible, nature also limits their functions; the laws of thermodynamics constrain the ways humans extract, transform, consume, and discard naturally produced matter and energy.

In order to understand how society and nature interact, and how the technologies that mediate between them emerge in social history but must conform to naturally regulated material laws and processes, we must resort to multiple human ways of knowing: in the physical sciences, in the social sciences—including economics—and in history. These multiple ways of social knowing and thinking are different from each other—and, indeed, cannot communicate directly—because they focus on three distinct spheres of understanding that appear to reflect three different kinds of reality: *physical and mathematical laws, material processes,* and *history.*

Physical and mathematical laws and rules apply universally. Most relevant to this study are space, time, gravity, inertia, momentum, the conservation and transformation of matter and energy, and the ratios between volume and surface. To the extent that these laws and rules are universal, they remain totally abstract and thus empirically unobservable. Humans have only approached some understanding of them when they are manifest in *material processes,* such as seasonal and diurnal cycles; the rate and timing of the growth and reproduction of plants and animals; the continuous cycles of evaporation, condensation, rainfall, and downward flow of water; the effects of water, wind, heat, and pressure on particular geological formations; the interaction of oxygen and flame; the interactions of different chemicals and minerals under different temperatures and pressures; or the transformation of gravity into various other forms of energy, including electricity.

Regular and controllable material processes are fundamental to economies of scale and diseconomies of space. Abstract physical and mathematical rules and laws configure differently to give them their various material and empirically observable manifestations. The gestation and life spans of a flea, a buffalo, and an elephant, for example, and the life cycles of trees and grass all obey the same physical laws and biological rules, but they vary because these laws and rules configure differently through the different material compositions of each species. Similar regularities and variations occur in geological formations, which result from varying configurations of universal laws through different material compositions, ranging from climatic patterns to the chemical reactions between different elements at different pressure and temperatures in the formation of rocks and minerals.

These second-realm material processes occur within first-realm time and space. We humans experience time and space only indirectly, through the regularities within the material processes that they govern. These regularities enable us to imagine what these laws are and how they operate, and we can constantly test our theories and beliefs about them through observation and measurement of the material processes within which they occur.

The diversity of matter—both local and global—results from spatially and temporally specific configurations of these unchanging physical laws and biological and chemical rules. This diversity creates the different technological potentials of different vegetables and minerals and of the spatially and temporally varied size and quality of their sources, which will be so important in our stories of how different nations developed new technologies. The regularities observable within each such configuration, and in the interactions between these configurations when they are conjoined or simultaneous, provide the potential for technological inventions. These technologies had to conform to, and depended on continuous supplies of, particular configurations whose technological utility defined them as raw materials, so these nations exploited, and then sought new and larger sources of, the raw materials that gave them the greatest technological advantage.

We must incorporate into our analysis the intersection of (a) time, space, and physical laws in the first realm, (b) specific material and energetic processes in the second realm, and (c) the technological discovery that occurs in the far less regular or predictable social history of the third realm. The third realm includes the diverse narratives—geological, biological, and social, in which we see the role of contingency, chance, purposive effort, competition, resistance, interpretation, communication, and ideology, that compose story, or *history*. This is where the inevitability of law and the regularity of process are joined with human choice, intention, and agency in the formation of society. It is also where the laws and rules configure as processes that create very different geological, climatological, and biological events according to the accumulation of site-specific material formations that we humans know as topography, vegetation, mineral composition, hydrology, and so on. It is in this realm that we can trace the development of technologies and the elaboration of national strategies for trade dominance that have driven the evolution of the world-system. These strategies only succeed, though, if they adapt to the material forms produced in geological or biological histories in specific locations.

These histories, like the material processes in which they unfold, occur in and are defined by time and space. Their outcomes are determined by the laws and processes that regulate their component ingredients. In the social realm, this means that technological discoveries and the search for the materials they transform will only succeed if they conform to these laws and processes; further, they will only succeed competitively if they conform better than alternative technologies already in use. The actual discovery, or the insight that leads to it, though, does not occur, nor can it be explained, through these laws and processes. Similarly, the elaboration of the political organization, the financial system, or the productive organization that enabled particular countries to become trade-dominant had to obey physical laws and rules in order to succeed but were not determined by them.

Obedience to physical law and advantage in material process are not sufficient to explain their success. Different human groups have different goals and interests. They have different intentions and are successful to different degrees in inventing technologies and strategies to achieve them. At the same time, all groups react to the technologies and strategies of their competitors, thereby changing the relative chances of success of everyone involved. Those nations that rose to dominate trade succeeded in part by anticipating or taking into account competitive or resistant behaviors both of their competitors and of their different suppliers. Their very success increased the complexity of the world context to which they had to adjust their strategies and actions. The required close-coupling and scope of coordination increased as the scale and scope of the world economy expanded and as the scale and cost of transport fell to bring more and more potential players into the same competitive game.

Thus, though each of these histories is conditioned by the advancing human understanding of and ability to use physical laws and rules, these cannot explain their formation, and can only explain part of their consequences. The other part of their formation and consequences is driven by the highly variable and necessarily contingent elements of choice, chance, skill, and information, as well as the path dependencies they form in the history of any human group's sequence of purposive actions and their outcomes. As technologies become more complex, more powerful, and more dependent on matter and energy from more diverse and far-flung parts of the globe, their effective use depends on the collaboration of the growing number of nations that control the places that have formed the different materials they consume.

Historically, one of the most effective and inexpensive ways to secure this collaboration is by interpreting and theorizing the history of the relations of particular societies to nature in ways that encourage other societies to treat their own local natural resources in ways that benefit the society that is reshaping the story. Industrial nations increasingly attempt to persuade the nations and firms that supply their raw materials that the resulting trade will benefit them. As the belief that industrialization resolves political and economic underdevelopment spread throughout the world, industrial states and firms and the intellectuals who serve them promulgated interpretations of their own histories that obscured the subordination of extraction to production. Some actually claimed that extraction was the source of national wealth. They incorporated these stories into economic theories that encouraged resource-rich nations to allow cheap and stable access to the raw materials in their territory.

These theories consistently overemphasize the role of capital, labor, and the importance of managerial and technical skill in its application while downplaying the absolutely essential base that land and raw materials provide for the effective deployment of capital and labor. Some of the most effective such mystifications are based on the synergies between mining, agriculture, the construction of vast railroad and steamboat networks, technological innovation, and industrial growth that made the nineteenth-century economy of the United States the most productive in the world. Some economic theories imply that these synergies resulted directly from general abstract laws, rather than from the particular national history that responded to the material features and processes of its own place and time. This is to confuse first-realm with third-realm realities. The illusion that the prosperity that productive economies reap from transforming natural products can also be reaped by the communities that extract or harvest them from nature facilitates the acquiescence of these communities to contracts obliging them to supply their raw materials at the lowest possible cost and thus to contribute to their own enduring underdevelopment.

When such mystifying interpretations are successful, they become hegemonic, achieving the complicity of the subordinate supplier nation with trade policies that benefit the industrial core while prejudicing the economy and environment of the extractive periphery. Core capital uses hegemonic discourses not just to legitimate and perpetuate this unequal exchange but also to convince local states to assume the costs of the large-scale extraction and transport systems that provide the cheap raw materials that core capital turns into profit. In the process, hegemonic discourse becomes mainstream theory—and hegemonic discourse is never more effective than when its proponents truly believe it.

We reincorporate these theories, and the national histories that form them, back into an analysis of how land, labor, and capital interact within all three realms. We use a three-level analysis of how land, labor, and capital interact to construct a materio-spatial logic. We then apply that logic across time to examine the ways that raw materials access strategies simultaneously expanded the domestic productivity of the ascendant economy, transformed international trade and investment, drove globalization, and exacerbated global inequalities—environmental and social. This allows us to compare local histories—first in the resource extractive periphery and then in the resource-importing core—to determine how successive struggles for world trade dominance between national economies generated technologies, financial institutions, and raw materials access strategies that enabled sequentially expanded economies of scale. We then consider how such increases in scale structured struggles over matter and space in resource-extractive economies, struggles that were organized in ever more distant peripheries to satisfy demands of core economies for raw materials appropriate to the technical and scale requirements of their expanding production systems. We also examine how each such step toward globalization exacerbated the inequalities between the consumers of socially produced commodities and the suppliers of naturally produced raw materials.

Assigning explanatory primacy to material processes as they intensify in expanded space does not mean that these processes are themselves independent of finance and politics, still less of the human intentions and culture that more fundamentally drive economic ambition. Rather, for reasons explicable only through an analysis of human or cultural traits as they have developed in specific places and times, different human communities have to different degrees been desirous of, and have to different degrees been successful in, the accumulation of material wealth and the subordination of other human communities in order to achieve it.

All human communities, however—especially those desirous of and successful in increasing wealth and conquest—depend upon material processes that manifest more general natural regularities. We humans understand some of these regularities as physical laws, others as chemical and biological rules; some of them we cannot pretend to understand at all. Historically, some of the more ambitious and aggressive human communities have learned how to manipulate and configure some of these material processes to their own advantage, sometimes through tools of warfare and conquest and, more effectively, through trade and transport. These efforts work to satisfy their ambitions and desires because trade and transport free them from the material limits of their immediate ecosystems. Dominance of trade and transport reduces the costs and enhances the benefits they receive from naturally produced raw materials. At the same time, it increases costs of and limits access to materials for their suppliers and their rivals by depleting or degrading natural production in their ecosystems in the short run and globally in the long run.

New technologies of extraction, transport, and production enhance the ability to transform a broader variety and greater volume of materials. New technologies require new forms of social organization, both in the dominant society and in those that it subordinates to expand its access to material. In this book, we follow the development of new technologies used to enhance the material wealth and political power of the communities that control them. We pay particular attention to the ways that these technologies must conform to the physical laws and to the chemical and biological rules that govern the material processes on which their functioning depends. We also follow the historical process of their sequentially cumulative increase in power, the ecological processes, and the consequences of the expansion of the variety and volumes of naturally produced material that the use of those processes consumes as the ways they serve human purpose increase in power, scale, and scope.

We also pay attention to the financial institutions and strategies that dominant societies devise to develop and implement these technologies and to the politics, both domestic and international, of achieving and enforcing the compliance of their raw material suppliers and their rivals within the systems of extraction, trade, and finance that these expanded technologies require. Finally, we examine how, historically, these dominant nations have reduced the cost of achieving acquiescence with social and economic structures that favor them by creating, divulging, and promoting economic and political theories, ideologies, and discourses that "prove" the material and the social advantages available to all communities that cooperate with them in their quest to procure and transform the natural products available in their particular ecosystems.

In this sense we consider the same data analyzed by conventional theorists of world-systems and hegemony who look to finance and politics, or to culture and communication, as the drivers of globalization. Like them, we theorize the dynamics of these socially constructed phenomena and processes, but, unlike them, we keep constantly in mind that, precisely because they are socially constructed, these dynamics are too susceptible to change, both intentional and unintended, manipulated or autonomous, to enable us to make law-like statements about them. We also keep constantly in mind that the greatest regularity manifest in these changeable social dynamics is that they achieve the human purposes for which they are designed and implemented only because of the opportunities afforded by the vast diversity of naturally produced material and spatial forms and, within that diversity, only to the extent that their technologies function within the boundaries of the materially possible. The laws and rules of nature that establish these boundaries are immutable, even if human technology contrives to combine and reconfigure the materials they regulate in ways and forms unlikely to occur in nature.

It is in the sense that nature provides both the material diversity and the absolute regularity of material reactions on which human technology depends that nature is primary, and, in Braudel's (1984) terms, has agency. Because humans increase the power and volume of technology and social production so rapidly, human agency and the third realm of reality that it shapes are highly apparent. More subtle, though far more important to all forms of life, are the far slower changes, over millennia, that differentiated primeval matter into the vast geo- and bio-diversity that sustain all forms of life and technology. We thus strive to remind the scholars who have so richly described the roles of finance, politics, and communication in the hegemonic transitions that have globalized the world economy that these powerful social processes are finally all about matter and energy, which are naturally produced and impervious to human agency, even if it seems as if they are primarily about money.

Increasingly powerful technologies have enabled the economies of scale that drive globalization. These technologies have demanded expanding volumes of a widening variety of raw materials. Different raw materials are differently produced in different places, so understanding the complex relations between matter, space, and technology requires detailed local histories of interactions between economy and ecology. Understanding how specific physical features and historic events observable in particular local places shape evolving global systems requires us to think about concrete local detail in ways that we can integrate with our more theoretical thinking about abstract global patterns and dynamics. The purpose of this book is to show how we can explain the interdependence between and the integration of the local and the global, the concrete and the abstract, and the social and the environmental. The stories we tell in this book will move from detailed description to abstract theory, but we will always return to the grounding of a specific place and time.

We will explain how matter, space, and technology, interacting in specific places through historical time, have shaped globalization. In the process, though, we will notice that the increasingly powerful technologies that have given humans increasingly effective tools to transform matter and space into useful, or profitable, commodities and infrastructure have also exacerbated social inequalities and environmental disruption. Technologies, and the economies of scale they make possible, provide dramatic benefits for, and impose dramatic costs on, society and nature. These costs and benefits are unequally distributed between classes, between regions, and between nations. The effects of both costs and benefits, and the inequality of their distribution, appear to become greater as technologies become more powerful and as economies become more global. Our purpose in trying to understand how local relations between space, matter, and technology interact with expanding global economies is to explain the history of raw material extraction in ways that will help civil society and political actors in resource-rich peripheries to resist the mystifying discourses of firms and states that seek cheap access to their raw materials, and to bargain instead for adequate rents, prices, and environmental protection. In order to do this, we will consider what actions those citizens of trade-dominant nations who hope to enhance social equality and maintain as much of the environment's integrity as possible can take toward those goals.

Sequence and Substance of Chapters

Our first work on Carajás aimed to analyze the mine's economic and environmental impacts. Bunker (1985) elaborated a model of the internal and external dynamics of extractive economies in the Amazon, with particular attention to the region's rubber boom. In 1984, he started tracing the complex struggles and negotiation over Carajás that led to the first shipment of ore in 1985 (Bunker 1989a, b). By 1989, Ciccantell had joined the project—first as a graduate student, then as a colleague.

We quickly learned that it was impossible to understand the mine without learning as well how firms and the state in Japan were working to shape the project in their own interests. As we worked to understand Japanese strategies for access to raw materials, we realized that they had been formed as part of a struggle against U.S. dominance of raw materials markets. As we compared U.S. and Japanese strategies, we realized that each trade-dominant economy over the preceding six hundred years had first achieved favored access to the raw materials most voluminously used in contemporary market production and trade.

This moved us to consider the sequence of extractive economies in the Amazon over those same centuries as a vantage point for analyzing the relationship between trade dominance and raw material access. As this work progressed, we saw that Carajás could only be understood in relation to (1) the history of global competition for trade dominance and (2) the history of local extraction as (3) the interaction between global and local as it changed through technological innovation. We learned that we could trace globalization through the case of Amazonian extraction.

Globalization is contentious because it tends to disrupt and devalue established social and environmental systems. Integrating socially intensified material processes across broader spaces requires huge investments in large-scale transport technologies. Transport systems open up previously isolated areas to the rapid social and ecological transformation, destruction, and conflict that occur on capitalism's frontiers. As extractive projects grow bigger and move farther from the industrial centers, they incorporate new zones, previously remote from capitalist production and exchange, into the world economy. New economies of scale imposed on hitherto unincorporated zones of the world-system disrupt established local ecosystems, increase the scale and concentration of capital, and lessen the authority and autonomy of the local political system. These effects leave both local society and local state with less competence and fewer resources to control the numerous problems of social welfare, social order, and environmental destruction that globally scaled projects with vast claims on locally produced matter bring with them.

The heavy infrastructure of globally scaled extraction, transport, and processing dramatically disrupts the physical environment, radically alters social and political relations, shrinks or eliminates established local economies, and subordinates local politics to global finance. These effects often generate intense conflicts between widely different groups that have differing amounts of power and operate in highly different domains, from global markets to local villages. The stakes are high for all of these groups. They range from huge capital, irretrievably sunk in remote places in order to secure cheap and stable access to raw materials critical to competition and profit at the global level, to the means of social organization, subsistence, and survival at the local level. Massive inequalities in power, scope of action, and organizational style often provoke different groups of actors to violent and destructive pursuit and defense of the matter and space they need to control in order to claim their stake.

Carajás's first shipment of iron ore from mine to port occurred only after complex political and financial struggles and strategies for designing, financing, and building the mine and its huge transport system provoked intense struggles over the spaces these projects opened up to capitalist economies. Carajás, and the Amazon rain forest into which the mine was inserted, are unique in both space and time. We believe, however, that the social, economic, and natural processes that created Carajás typify systemic relations between the numerous heterogeneous extractive economies that supply the raw materials consumed by modern industry and a far smaller number of advanced and increasingly homogeneous industrial economies that transform and consume them. The technologies that determine which materials industrial societies consume, and on what scale and volume, though, change over time, so the raw materials most important to trade dominance and the relative advantage of different nations in access to critical raw materials also change.

In chapter 2, therefore, we consider how the raw materials exported from the Amazon changed in ways that reflected the technological changes in the world economy. The physical and topographical features of these different raw materials shaped their extraction and transport, the labor and property relations involved, and the environmental impacts caused. The Amazon provided raw materials for the dominant technologies in each of the world's successive regimes of trade dominance. We will show how both the causes and the consequences of the sequence of resource-extractive economies in the Amazon resulted from interactions of natural and social dynamics at the local and the global levels.

The rest of the book uses what we have learned about extractive economies to analyze the strategies of productive economies to dominate trade. Chapter 3 examines and compares the competitive strategies and international relations of each trade-dominant national economy of the past five hundred years. Chapters 4, 5, and 6 consider the relations between the states, firms, and sectors in each of the countries that achieved trade dominance since the original colonization of the Amazon—by Portugal, Holland, Great Britain, the United States, and Japan. Chapter 7 synthesizes these historical and physical comparisons and considers what we, as scholars and citizens of local and of global systems, can do to slow or reverse the environmental destruction, the social disruption, and uneven development that result from the expanding scale of economies. As we tell each separate story, we will refer forward and backward to the stories of other trade-dominant nations and of their peripheral suppliers in order to analyze persistent patterns and dynamics that drive change in the entire world-system.

CHAPTER TWO

Globalizing Economies of Scale in the Sequence of Amazonian Extractive Systems

Technologies, commodities, and markets must conform to the physical and chemical features of the materials they transform and consume in order to provide the desired goods, services, and profits. Extraction and transport of these materials, in turn, must conform to the topographical, geological, and climatological features of the many different places where the materials are naturally produced and to the space between where nature produces them and where humans transform, exchange, and consume them. The spatially differentiated physical and chemical features of matter constrain what kinds of technologies can produce what kinds of commodities at what kinds of prices, so competitive capitalism generates complex interplays between technological and commodity innovation in the industrially dominant core and space and matter in the extractive periphery.

One way to observe and understand these interplays is to chart historical changes in the kinds and volumes of raw materials taken from a specific region as innovations in technology and in commodities require new and different raw materials. In this chapter, we trace the sequence of extractive economies in the Amazon basin over the four hundred years since its first European colonization. During those four hundred years, the Portuguese, then the Dutch, followed by the British, the Americans, and finally the Japanese, devised new technologies and markets to achieve world trade dominance. These innovations changed the kinds and volumes of material that each national economy demanded from the Amazon.

The material and spatial features of the Amazon, though, set the physical, financial, and social-organizational conditions to which those technologies and markets had to conform in order to serve the purposes of the capitalists and of the states that mobilized them. Human social organization must obey the physical, chemical, and biological laws, rules, and regularities that configure to create the distinctive environments from which humans extract material resources. In this sense, the natural environment is more powerful than the human societies that depend on it for subsistence or profit.

Paradoxically, though, the reproductive dynamics of the coevolved and complexly interdependent material and energetic systems that make up most environments are extremely vulnerable to commercially generated disturbance. Overharvest or depletion of a single species can ramify through multiple species in long chains of dependence. These problems are self-limiting in most subsistence economies, but they can become severe when trade across ecosystems generates regional specialization in particular extractive goods.

Humans who subsist entirely from one ecosystem must satisfy their multiple needs from a multiplicity of locally produced materials and so spread the environmental impact of their harvest across many different species and land forms. Harvesting more than they can consume or use of a particular material would be a waste of their energies. If they do start to overharvest a particular material form, their harvest will cost even more energy as that form becomes scarcer. To the extent that overharvesting a particular form reduces the entire ecosystem's carrying capacity, it ultimately leads to human population loss and to a reduction of the harvest. Most human societies dependent on a single ecosystem learn to adjust their activities, consumption, and often their own reproduction in ways that sustain both the particular environment on which they depend and their own population densities.

Humans who trade across ecosystemic boundaries, in contrast, tend to focus their extractive activities on the few resources in any particular environment that will return the most profit to their efforts and to extract more of these resources than the local human population can use. Extractive economies thus often deplete or seriously reduce plants or animals, and they disrupt and degrade hydrological systems and geological formations. In complex, bio-diverse ecosystems, these species and topographical features participate in complex chains of material and energetic exchange. They serve critical functions for the reproduction of other species and for the conservation of the watercourses and land forms on which they depend. Losses from excessive harvesting of a single species or material form can thus ramify through and reduce the productivity and integrity of an entire ecosystem. The Amazon, the most complex and bio-diverse system on the planet, is no exception; each extractive cycle there changed the local environment and thus changed the conditions for subsequent extractive economies and for contemporary subsistence.

In each extractive cycle, the initial loss of the targeted material form ramified through a web of coevolved and thus interdependent species and through the land and water forms that sustain their interaction. The physical, chemical, and biological mechanisms that configure Amazonian ecosystems continue to operate independent of human activity, but the material and spatial forms they configure, and thus the environment they create, are highly susceptible to human disturbance. To be successful, human extraction from these systems must adapt to the natural processes that create and sustain them. In each case, though, the initial success of human extraction from these systems triggered a series of natural reactions that disrupted naturally evolved flows of matter and energy, reducing the environment's productivity in unintended and unanticipated ways. These losses impoverished the human societies that drew their subsistence from that environment, and they hastened the depletion of the resources being extracted from it for export. In the rest of this chapter, we will consider the implications of these and other specific examples of extractive economies for the general model we developed in the last chapter.

Matter, Space, and Labor in the Decimation of the Amazon's Indigenous Populations

The Portuguese settled and fortified the Amazon in the early seventeenth century. In order to cheapen its occupation and administration of the Amazon, the crown granted its officers huge tracts of land and rights over indigenous labor in the territory it had just appropriated. The devastating effects of their efforts to convert local biotic forms into commodities for export to global markets epitomize the ways that extractive export economies degrade nature and society.

The Amazonian space was not appropriate biologically, hydrologically, or climatologically either for the technologies or for the labor relations preferred by the Portuguese. Portuguese attempts to grow sugar in the Amazon failed. Their subsequent attempts to exploit the Amazon turned, instead, to extracting luxury goods from the exuberant tropical environment that surrounded their forts. These extractions were never profitable enough to purchase or sustain the imported slave labor that sugar supported farther south. Instead, the Portuguese enslaved the indigenous population. They exploited the captured labor to construct edifices and roads; to extract cacao, rosewood, and spices from the forest; to hunt manatees, capybara, and caymans; and to gather the eggs that turtles had laid in nests along the rivers.

These commodities all had very high value-to-volume ratios in European markets, but they were broadly dispersed in very small amounts across complex, vegetatively dense, frequently flooded spaces that made their collection highly labor-intensive and environmentally destructive. Turtle oil, for example, was much prized by Europe's wealthy as a perfume and as fuel for their lamps. One cup of this luxury liquid required finding hundreds of turtle eggs, smashing them into a canoe, and then skimming off the thin layer of oil after it rose to the surface of the liquid mess. This wasteful harvest decimated the turtle population, reducing the indigenous population's access to protein from turtle meat and from fish that preyed on baby turtles. Costs rose and profits fell as Portuguese expeditions organized for slaving and for extraction depleted the most proximate sources of labor, meaning that they had to travel farther and farther upriver. Like all extractive export economies, this one was directly impoverishing itself.

Indeed, the Portuguese conducted their slave raids as if they were extractive enterprises—that is, as if the Native Americans were forces of production that they could appropriate from nature. Because indigenous societies had adapted their subsistence to the ecosystems that the Portuguese were harvesting for commercial profits, the depletion of the extracted resources paralleled the decimation of indigenous populations. In the rest of this section, we explain how the depletion of resources and the decimation of human populations reinforced and accelerated each other, impoverishing the environment, local social organization, and the colonial economy in self-aggravating cycles.

The conditions of work and life in an increasingly impoverished settlement, combined with exposure to exotic germs brought in on ships from Europe, caused numerous epidemics and the consequent reduction of enslaved labor.

The need to replenish their supply of slaves drove the Portuguese to mount more slaving raids and to provoke slaving wars, which drove the remaining indigenous population farther and farther upriver, increasing the distance to be traveled on subsequent raids. Longer distances required more provisions—for Native American slave rowers on both legs of the journey and for newly captured Native American slaves on the return. The increasingly impoverished colony did not have the resources for additional provisions, so slavers tried to save capital and cargo space by reducing the food they carried with them. Both rowers and new captives suffered malnutrition, disease, and death as a result of the slavers' desperate attempts to economize transport across ever-greater spaces (Hemming 1978; Sweet 1974).

Indigenous adaptation to their aquatic environment increased their vulnerability to capture and disruption. The densest populations had located around the mouthbays of tributaries to the Amazon, where the calmer waters provided access to fish and turtle protein. Yearly floods of up to thirty meters maintained soil fertility and reduced pest populations, but they also imposed the need for mobile and versatile subsistence strategies. Highly complex and very productive configurations of space and of matter in and around the mouthbays had molded, facilitated, and enhanced indigenous techniques that combined hunting, gathering, fishing, cultivation, and transit in this biodiverse, seasonally variable ecosystem. Indigenous groups incorporated complex knowledge of plants, animals, and bodies of water into social organization, cosmology, and technology.

The mouthbays and *varzeas*—spaces with abundant fertile grounds to cultivate and waters to fish—became dangerously accessible to Portuguese boats. Portuguese slaving drove the populations that survived war, disease, and enslavement up the tributary rivers to the less fertile land and the less rich waters of the *terra firme*. The technologies devised for the *varzea* were irrelevant there and consequently were lost (Ross 1978), both to the Native Americans and to their Portuguese conquerors. Portuguese extraction impoverished both the environment, on whose resources the indigenous populations depended, and the knowledge of how to use those resources. Enslavement, disease, war, flight, and refuge in less fertile environments hugely reduced indigenous populations.

The same attributes of space and matter that had molded indigenous knowledge and social organization constrained Portuguese attempts to establish techniques of agriculture and trade that they had imported from an altogether different environment. They also structured Portuguese violence and exploitation and set the patterns for indigenous reaction and flight. The inter-action between these natural and social forces created a demographic vacuum that seriously impeded local response during the technological and market changes that were brought on and then sustained by the nineteenth century industrial revolutions in Europe and in North America, which created rapidly growing demand for Amazonian rubber. The new machines of rapidly ex-panding core industries tremendously accelerated the consumption of raw materials. Expanded extraction and trade radically altered the societies and environments of the peripheries that provided raw materials, but the config-uration of material and space in those peripheries shaped the world economy that consumed them. In the next section, we consider what happened in the Amazon when the industrial boom created by iron and coal in Europe and in the United States fostered a demand for far more rubber than natural sources could supply.

Property, Labor, and Exchange in the Rubber Economy

The second industrial revolution created an insatiable demand for rubber. European inventors, engineers, and capitalists had found ways of combining iron, coal, and steam that enabled them to build and power machines that performed an increasingly broad range of work. These machines could move far greater loads more rapidly and apply far greater force at much higher temperatures than either human or animal effort could manage. Innovation and capital rapidly extended these functions to the development of machines that made other machines. These innovations and inventions stimulated the discovery and use of other raw materials, such as aluminum, cassiterite, and manganese, and other energy sources, such as petroleum. Further innovation adapted machines to the physical and chemical properties of the new types of matter discovered or processed.

Intense demand for these materials often stimulated local economic booms and environmental and social disruptions in the places from which it was most convenient to extract them. Amazonian rubber was one such material. The critical functions it served in a broad range of new technologies and products catalyzed a major economic boom in the Amazon, soon to be followed by major environmental and social disturbances.

New technologies to process rubber of a quality found only in the Amazon made it possible to stabilize and toughen rubber against stretching under pressure and heat while allowing it to maintain its inherent flexibility. Rubber thus vulcanized could be made into belts that transmitted energy from increasingly powerful steam engines, then into electric motors and internal combustion motors, and finally into a widening range of machines in industry, agriculture, and extraction. Vulcanized rubber could be made into pads for moving parts that rubbed against each other, insulation for metal that conducted electricity, and tires for the wheels that made machines mobile. The firms that built or used these machines realized huge and expanding surplus profits, the national economies in which these firms participated realized rapid economic growth, and the states that directed these national economies greatly increased their revenues and powers.

Many of the new technologies were first discovered by military inventors searching for deadlier and more efficient tools of violence. Control over the raw materials that made up and powered these machines, and over innovations that extended their productive or violent use, directly affected the security, prosperity, profits, and survival of states and firms alike. Therefore, they tended to collaborate in a broad range of strategies and campaigns to assure cheap, stable access to adequate quantities and qualities of these raw materials.

Because rubber performed such crucial mechanical functions, and because the Amazon provided the technically most useful and abundant kinds of rubber in the world, states and firms in the rapidly industrializing, trade-dominant nations of Europe and America were particularly interested and active in the Amazon. The size and climate of the Amazonian territory, the extent of its rivers, the demographic distribution and social organization of its human populations, and the politics of the region and of the nation that claimed to control it all set complex conditions to which foreign capital and states had to adapt if they were to gain stable access to the rubber they needed.

As steam-driven pumps and motors mechanized world capitalism, their importance to business profits, to state revenues, and to military might stimulated searches for technologies that stabilized rubber even at high temperatures and levels of friction. Charles Goodyear's 1839 invention of vulcanization permitted the development of flexible belts to enhance the transmission of mechanical energy and of tires to allow mobility for these machines. Subsequent research and investments improved vulcanization, improvements that progressively extended the profitable mechanical applications of rubber.

The rapid expansion of the mechanical uses of steel and the rapid development of new mechanical technologies eventually generated incentives for the invention, fabrication, and standardization of machine-tooled screws, cogs, wheels, and shafts. Once these new inputs were possible, the steel from which they were made transmitted more force, resisted greater heat, and lasted longer than rubber. At the beginning of the machine age, however, rubber's material qualities—whether natural or artificially improved—allowed less technically demanding performance of these highly profitable functions. As steel quality improved and unit costs dropped, the market for metal machines that used rubber—either as tires to permit movement on wheels or as belts to transmit power from one plane (e.g., horizontal) or motion (e.g., rotary) to another (e.g., vertical or linear)—soared, driving the booming demand for rubber.*

Rubber of a quality found only in the Amazon worked best for these new technologies. Much of the most profitable research and development of new technologies were thus based on Amazonian rubber. Incremental refinements in technology combined with inventions of new mechanical applications to drive demand for rubber to levels far higher than local labor and transport systems could meet. Prices soared, and the local attempts to respond to this booming demand radically changed the economy, the demography, the politics, and the law in Amazonia. Material and spatial features of the Amazon's environment, together with the enduring consequences—demographic, ecological, political, and economic—of the earlier European exploitation of its human population and of its natural resources, made it impossible to satisfy soaring demand. Local attempts to corner or manipulate the market exacerbated the volatility of already climbing prices and the irregularity of the already tenuous and inadequate supply.

The rubber boom responded to the demand for specific material functions required by the new technologies that drove the second industrial revolution— one of the most dynamic episodes of material intensification and spatial expansion in history. The discovery of vulcanization and the rapid proliferation of new mechanical uses for rubber closely paralleled the development and spread of the process of Bessemer conversion for smelting iron, which made durable steel of uniform quality cheap enough for mass production. Bessemer steel made possible the rapid mechanization of agriculture, extraction, and industry as well as the cheap and rapid transport of the bulky raw materials

---

*A partial list of such machines would include bicycles, automobiles, tractors, reapers, combines, threshers, silo loaders, saw mills, conveyor belts, electric generators, and steam shovels, not to mention mechanized flourmills, bakeries, sewing machines, and even chocolate makers, and, less profitably, tanks, jeeps, and machine guns.

that drove the rapid ascent of the U.S. economy in the second half of the nineteenth century.

The new steel-based technologies that depended on rubber provided huge surplus profits and stimulated major economic growth in the rapidly industrializing core, particularly in Britain and the United States. Industrial firms and national states were intensely interested in the supply and price of rubber, but they had to adapt to the materio-spatial, economic, and political structures of the Amazon to satisfy their needs.

The local actors who organized to take advantage of the new demand for rubber were directly constrained by (1) the biological characteristics of the rubber tree, *hevea brasiliensis,* particularly as those characteristics determined the spatial distribution of the tree itself and the temporal distribution of the labor process involved in tapping, collecting, and curing the rubber, (2) the course and flow of the rivers that provided the only commercially viable access to the *seringais,* or rubber groves, and (3) the seasonal patterns of rainfall and flooding.

The biology of the trees, and their distribution in space, had evolved to a very wide dispersion of single trees across spaces broad enough to impede the proliferation of *dothidella ulei,* a fungus that thrives where rubber trees form thick groves. A tapped rubber tree drips latex slowly enough that a cup could be collected only every second day. Gathering rubber thus required single individuals to walk great distances between trees to collect a relatively small amount of latex rubber each day. They then had to cure the latex and agglomerate it in small daily increments into a large ball of rubber for eventual delivery against their debts to the capitalist who had transported and provisioned them.

Just as the high cost and slow speed of animal traction limited seventeenth-century logging described by Albion (1926) to trees closer than three miles to a riverbank, the time and effort of carrying latex through the jungle limited the paths, or *veredas,* to distances within the area that a man could tap and carry to a riverbank in a day's work. The inward transport of labor and provisions at the beginning of the dry season and the outward transport of labor and rubber at the beginning of the rainy season depended completely on the course and flow of rivers, so rivers determined which *seringais* could profitably be tapped. The low density of rubber trees and the restriction of tapping to trees within a day's round-trip circuit from a river meant that response to increasing demand required longer voyages upriver. This increased labor and transport needs and so drove up the cost and price of rubber.

The demographic vacuum left by the earlier enslavement of indigenous populations imposed the need to transport labor and then to control and discipline that labor across space. The absence of local populations near the rubber groves increased the cost and complexity of the rubber merchants' adaptations to the biology of the trees and to the flow of the rivers that provided avenues of access. The merchants had to advance capital sufficient to sustain isolated rubber tappers long enough to accommodate the slow, biologically determined rhythms of latex dripping. The tapper had to smoke and agglomerate the small amounts of latex he gathered each day into slowly accumulated balls large enough for cost-effective transport to and handling in international raw materials markets. Direct control of labor, direct security over the capital advanced for its transport and provisioning, and physical security over its product under these conditions were all impossible.

Time, space, and matter thus all threatened the capitalist's ability to ensure that he could appropriate rubber as return and profit on his investment. Capital was impelled to organize the labor process through devices that guaranteed returns on investments in transporting, sustaining, and disciplining isolated laborers at great distance over extended periods of time.

The rubber traders, or *seringalistas*, devised customs, and the state devised legal forms of property and usufruct, based on the flow of rivers that designated the owner or lessee of a *seringal* as the sole legitimate claimant to rights of transport on each river. Rubber-trading firms organized hierarchies of traders to match the tributary systems of the major rivers that provided access to the lands they controlled, so that subordinate traders were each in charge of gathering rubber from, and preventing rival access to, particular tributaries of larger rivers. The firms themselves received supplies and capital advanced by international rubber companies as debt, supplies and capital they distributed upriver to the subordinate traders on credit, who advanced them as debt to the tappers even farther upriver. The rubber then flowed downriver in a reverse trajectory, to pay the debts contracted earlier at each level of the hierarchy.

The natural material and spatial configuration of rubber trees and rivers thus created the parameters within which social, economic, and political relations of property, extraction, transport, exchange, and law could be arranged to fit both the global economic demand and the local situation in ecological and social terms. Capital and supplies imported by British and U.S. firms and advanced to the *casas de aviamento* sustained the debt and violence-driven *aviamento* system that evolved to fit this confluence of natural resource, market opportunities, physical constraints, and capitalist intent.* The peculiarities and importance of local social and political custom, law, and knowledge forced international capital to rely on legally autonomous local merchants for its material operation (Santos 1968, 1980; Weinstein 1983).

The local state and various local business associations attempted to encourage investment of rubber trade revenues into other local industries and agriculture, but as long as the market for rubber offered far higher profits than alternative economies in the Amazon, capital was strongly induced to continue investing in rubber and the state in storage and transport infrastructure. Besides having to compete with the far more attractive investment opportunities offered by the rubber trade, local industry also had to compete with commodities imported very cheaply from Europe and the United States in the ample empty cargo space on the return leg of the ships that carried the huge quantities of rubber to economies where consumer goods and agricultural equipment could be produced more cheaply than in the Amazon. Cheap transport for imports and the strong demand for labor in the rubber trade actually shrank production in local agriculture and industry, even though the amount of money circulating in the local economy, and therefore the markets for food and commodities, grew enormously.†

Limited supplies of rubber created serious bottlenecks in multiple, and potentially highly profitable, industrial sectors. The potential for profit in rubber-dependent technologies accelerated even more with the invention and spread first of bicycles and then of automobiles—the first mass-produced machines produced as end-use consumer items rather than as means to produce other commodities. Markets for both of these machines exploded, generating new mass production techniques and huge surplus profits. Technical and product innovations and the rapid mechanization of military force added to the hugely

---

*The system took its name from *aviamento,* the Portuguese word for "providing." Merchant capitalists in this system were known as *aviadores.* The rubber grew in *seringais* leased in long-term *aforamentos* to *seringalistas.* The *seringueiros* delivered the rubber they tapped and cured to pay off the debt for transport and provisions. The *seringalista* or the *aviador,* who might or might not be the same person, set the prices for provisions and transport and for the rubber that was delivered against them. Tappers' debts thus tended to be perpetuated across seasons.

†Coronil (1997), noting a similar set of pathologies in Venezuela's oil exporting economy, points out that the problem is older than the 1970s crisis that natural gas exports created in Holland which resulted in the term *Dutch disease.* He suggests the problem should be called *Third World disease.* The booming market for a tropical raw material from the Amazon created economic problems very similar to those that Innis (1933, 1956) identified in Canadian economies set up to supply markets for sub-Arctic raw materials.

expanding demand for rubber. Unmet demand drove rubber prices higher; local speculation and attempts to corner the market made these prices volatile around a very high base. The spatial and material limits on rubber extraction in the Amazon directly limited global supplies as industrializing nations became increasingly dependent on and enriched by rubber's role in production.

These conditions inspired the British state to botanical, political, diplomatic, and economic efforts to domesticate the rubber cultivar. British science joined with British imperialism and transformed rubber from a wild plant in Latin America, where British capital "controlled neither land nor labor," to a plantation crop in Asia, where British capital controlled both (Brockway 1979).

The flood of plantation rubber onto world markets in 1910 reduced the price of rubber below the costs of extracting, transporting, and exporting feral rubber from the Amazon. The ports and boats of the *aviamento* system were too costly to maintain under this new price regime. Most were allowed to deteriorate while their diminished use values were applied to a vastly reduced trade in rubber, gathered now by autonomous peasants as a supplement to diversified subsistence activities.

The successful transformation of rubber from limited feral production to expansible domestic plantation removed the constraints that the materio-spatial features of the Amazonian *seringais* had imposed on expanded production and accumulation in the industrial core. This transformation ultimately impoverished the Amazon's economy as rapidly and radically as local responses to growing core demand had initially enriched it. Space and matter defined the mode by which the Amazon was reincorporated into the world-system as the source of an industrially critical raw material. At the same time, space and matter created the conditions that eventually led coalitions of firms and state agencies, in collaboration with colonial offices and scientific organizations, to search for technological, organizational, and imperially coercive means to increase and stabilize the supply and to reduce the cost of rubber. Agents of industrial firms and of imperial states collaborated to move the reproduction of rubber to new locations and to import colonial populations for its cultivation there.

How Matter and Space Mold the Technologies that Drive Globalization

Naturally produced materials become economic resources only as humans discover or invent uses for them, access to them, and transport for them; but human endeavors to exploit the resources produced in specific locations can only succeed if they adapt to the material and spatial features of those locations. Each of the Amazon's extractive cycles—the slave wars; the trade in turtles, fish, cayman, manatee, and capybara; the rubber boom and its eventual decline; and the contemporary iron mine and aluminum complex—was catalyzed and constrained by the technologies of production and transport and by the consumer markets they served. Physical and chemical properties of each of these resources determined the human uses they could serve and the technical processes required to transform them into useful forms; access to each resource provided opportunities for surplus profits to the capital that developed and controlled the technologies to transform them into commodities. These opportunities motivated agents of the state, of finance, and of firms in the exporting and importing economies to apply considerable capital, energy, and political power to assure cheap and stable access to these resources.

In each extractive cycle, the river system as a hydrologically, topographically, materially, and biologically differentiated space created the conditions for the reproduction of the resource extracted and the means of its transport. The biology of rubber trees, for example, determines both the material properties that make rubber and the distribution in space that once made it difficult and costly to extract and export, while the hydrology and the topography of the rivers determined the relative costs of exploiting differently located groves. The cost of extraction determined how much of the limited maximum possible supply of Amazonian latex actually reached the market as rubber.

The successful British efforts to transform rubber from an extracted feral resource into a cultivated crop hugely increased world supplies and greatly reduced world prices of rubber (fig. 2.1). The technological and economic effects of this increase illustrate the ways that natural materio-spatial systems constrain and mold socioeconomic systems. As long as world supplies depended on extraction from the Amazon, limited supplies and high cost retarded the technological impulses of the dynamically expanding automobile, shipbuilding, and machine-tooling industries. World consumption of rubber increased fivefold during the decade following the 1910 introduction of plantation rubber on international markets. This expanded, cheapened supply enabled the development and proliferation of technological innovations that incorporated rubber into an expanding range of increasingly productive and profit-generating industrial machines and, during World War I, of increasingly destructive military machines. These effects of the domestication of rubber underscore the specific effects of local biology and hydrology on the

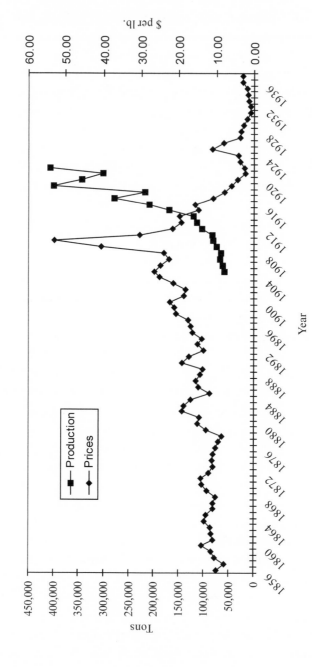

Figure 2.1. World Rubber Production and Prices, 1856–1937.
Note: Rubber prices are for wild rubber, FOB in New York City, converted to 2000 US$. Production is for both wild and domestic rubber.
Sources: Barker (1938); U.S. Dept. of Commerce (1925).

global economy and the more general constraints of matter and space on technology.

The materio-spatial attributes of rubber and of the river system that conditioned production and commercial access to it affected the economic opportunities and the social and political organization of firms, financial institutions, and state agencies in the industrializing nations as much as it affected merchants, rubber tappers, boat owners, and politicians in the Amazon. The ways they affected these different social groups, however, and each group's behavior in response, varied enormously according to their different positions in relation to the materio-spatial characteristics and to the technical and economic processes to which these characteristics gave rise. The consequences of these actions—their costs and their benefits—were unevenly distributed. The sequentially cumulative processes of technological and social-organizational innovations in extraction, transport, and processing emerge from interactive dynamics between local and global instances of nature and society.

The dynamics change as technological innovation, driven by competition for trade dominance, changes industrial needs and reduces the cost per ton-mile of transporting material to supply them. The material and spatial properties of these resources are set naturally; the technical, financial, political, and organizational devices for their extraction, transport, processing, and exchange are created socially.

These reciprocal dynamics belie notions of the social construction of nature and of space. Humans do not transform or construct space socially; at most they accelerate and intensify their own transit and their transport of matter across it. Time does not annihilate space in this model, rather space as a condition of production provides opportunities, and space as an obstacle to transport presents challenges that foment innovations in social organization and technology. Such innovations can only reduce the cost of transport by increasing the scale and speed of moving matter through space. Overcoming the increased physical stresses of weight, friction, inertia, momentum, and impact that such economies of scale and speed entail requires new, more powerful machines made from new and stronger materials.

That local processes in the Amazon produce global consequences demonstrates that the origins of global processes are rooted in locally observable and specifiable materio-spatial features. New technologies of increased scale and speed create new needs for more, different, and stronger kinds of raw material than can be found in any one of many spatially distinct local systems. The diversity of matter in space increases the cost of transport, but combining and coordinating these different materials enables the economies of scale in transport and in technology that have driven the centuries-old progress of globalization. The technological, social, organizational, and financial innovations that accommodate space and matter to human purposes and mediate the contradiction between the cost of distance and economies of scale thus drive the progressive globalization of capitalism. The timing of these innovations both drives and reflects the material intensification and the spatial expansion of the world capitalist economy, as do their sequentially cumulative scale, productivity, and profitability.

The sequence of new technologies intersects and interacts with material resource discoveries in particular spaces. These discoveries often motivate or accelerate the incorporation of previously external areas as peripheries into the world-system. Local actors—sometimes indigenous or previously resident but often immigrants responding to the economic opportunities the recently discovered resource provides—drive political, social, economic, and financial reorganization in their endeavors to profit from changing demands. Their profits depend on adapting to and exploiting the materio-spatial features of the resources in ways that satisfy the technical needs of the processes of production and the financial needs of the states and firms that control these techniques.

Eventually, socially produced global demand overshoots naturally produced local supply. Local elite responses to the pressures and opportunities that the expansive and technically driven world-systemic demand for raw materials creates eventually change the ecology that they are exploiting, generally reducing its capacity to keep up with that expanding demand. Economies of scale in social systems of production can continue to expand as long as (1) market demand continues to grow and (2) capital and labor can continue to import and transform increasing volumes of more types of raw materials. Raw materials, however, must be extracted from natural systems of production. Social systems of production can achieve significant economies of scale and speed, especially when they are based on mineral rather than organic materials. Natural production resists increased scale and speed. It is diffuse across space in ecologically distinct places. Vegetable production depends on horizontal space for capture of solar energy and terrestrial minerals; vegetable growth is subject to biological and seasonal rhythms that cannot be accelerated. Mineral production can accumulate vertically as seismic or hydrothermal movements open and fill subsurface cavities with magma or as decomposition, filtration, and sedimentation deposit mineral matter in successive strata, but both of these processes happen in geologically determined places and times. The accumulation of deposits adequate for commercial extraction is both rare and very slow. Industrial production, in contrast, can agglomerate in places homogenized by social groups that transform their physical environment to expand, accelerate, and cheapen their transformation of matter into commodities and accelerate the turnover time of capital. Natural systems of production impose diseconomies of scale when social systems of extraction are organized and accelerated to meet the growing demand created by economies of scale in industrial systems. Local responses to increased demand deplete mineral sources and overharvest organic sources in particular places.

The empirical narratives of the boom and bust of each of the Amazon's extractive episodes reveal the materio-spatial consequences of depletion and of the diseconomies of scale in most extraction. Taken together, these narratives suggest that when the lag between global demand and local supply becomes too great, states and firms in the core mobilize science, commerce, imperialist or colonial forces, and debt finance to resolve problems of rising cost, scarcity, or inconsistent supply by (1) finding sources of the raw material in question in other locations (and arranging transport systems to accommodate them), (2) finding natural or devising technical or synthetic substitutes for the raw material in question, or (3) domesticating and converting the natural or feral sources of the raw material to plantation cultivation, which necessitates importing or enslaving a population sufficiently controllable to provide a cheap and stable labor force.

All three of these solutions manipulate material process, both natural and social. They relocate sources according to social (i.e., demographic, political, and geopolitical) and natural (i.e., climatological, topographical, and geological) attributes as these relate to the material requirements of the reproduction and extraction of the raw material and to the spatial, topographical, and hydrological determinants of the cost of access and transport. All three of these solutions provide mechanisms to explain the spatial expansion and material intensification of world-systemic production and exchange that is now called globalization. All three modify the binary and multilateral production and exchange relations that structure the world economy. Each of these alternatives materially intensifies and spatially expands the world economy while devaluing the capital the local economy has sunk into adapting itself and its environment to accommodate the world-system's needs for its resources.

The physical and chemical attributes of raw materials and their location in space as mediated by topography, hydrology, geology, climate, and biology directly constrain and shape all three of these social and geopolitical strategies for the extraction, transport, transformation, exchange, and consumption of the expanding diversity and volume of commodities. The deployment of these strategies intersects with the economics and geopolitics of space. This intersection implies that materio-spatial mechanisms both shape economic strategies and underlie the ways that these strategies mold social structure, agency, and organization from local to global levels.

Appropriate comparisons to other site-specific conjunctions of global demand with the spatial and material configurations that produce particular kinds and quantities of natural resources help us to abstract natural physical (as well as social, political, and economic) mechanisms from these narratives. We can then use these mechanisms to typify general relations between the increasingly homogeneous, progressively globalizing industrial core and the increasingly heterogeneous extractive peripheries that export just one or a few of core industry's multiple, expanding, and still diversifying material needs.

Analyzing each of the multiple spatially and materially distinct local economies requires close attention to extensive detail and tends to result in highly descriptive, or ideographic, accounts. Analyzing and theorizing how material processes within expanding spatial relations globalize the world-system require a much more abstract, unitary, and systemic, or nomothetic, analysis. Integrating local and global analysis thus poses a very thorny set of problems. The vast number of such peripheries militates against comprehensive comparative analysis. The variation and complexity in the ways that matter, space, time, technology, finance, politics, and social organization configure in each locale require highly detailed analysis of multiple interactions rather than a panel analysis of a limited set of standardized variables.

Typification from a particularly salient set of local materio-spatial characteristics as they interact with changing patterns of globalizing capital over centuries provides a promising approach to these problems. Typification from a single case to the complex intersection of many heterogeneous cases within a global system tending toward increased homogeneity requires that we interpret the single local case through powerful theories adequate to account for the dynamic of the global system. In the next section, we demonstrate how this can be done for one of the most significant material transitions in history—the shift from primary reliance on organic matter to primary reliance on minerals.

The Transition from Organic to Mineral Raw Materials

Locally dominant classes organized each of the Amazon's extractive economies to exploit new markets for the resources available there. The collapse of the rubber economy dramatically reduced local connections to world markets for thirty years. After 1950, though, the Amazon was reincorporated into the world-system as a supplier of industrially critical minerals on a financial and technological scale that far surpassed that of the rubber trade. Paradoxically, though, as financial and technological scales increased, the Brazilian state, large Brazilian companies, and the state of Pará, all rooted in the periphery, were encouraged to assume a greater share of the management—and a greater share of the risk—of these enormously complex and costly projects. This devolution of risk and cost, obscured by hegemonic—and mystifying—discourses that promise local development from nationally controlled extraction, has been an increasingly important component of the competitive strategies core nations use to drive globalization.

In the sequence of Amazonian extractive cycles we can discern the trajectory of international military struggles, economic competition, and innovation in the technologies of production and transport at the core of the world economy. These same core processes shaped each of the systemic cycles of accumulation that occurred during the four centuries of Brazil's incorporation into the world-system, so the Amazon's raw materials exports provide a peripheral perspective on how world-system dynamics drove globalization. Struggle with the Dutch for dominance of the sugar trade and over commerce in iron tools with indigenous groups (Sweet 1974; Bunker 1985) motivated Portuguese military fortification and slave raids at the time that sugar was becoming the most valuable crop in international trade and that the Dutch were augmenting their dominance of world transport with high-quality industrial fabrication of metals. Military authorities and Jesuit missionaries organized the destructive exploitation of turtles to profit from luxury demand in Europe for combustible oils and for self-preserving meat that could be used on long wind-driven ocean voyages. These high-value/low-volume commodities were typical of long-distance trade before industrial technologies made it possible to convert petroleum into light, heat, refrigeration, and motor power.

The rubber economy's rapid rise and vertiginous fall were driven by the accelerating cycles of technological, scientific, and organizational innovations of the second industrial revolution, which first created and expanded Britain's industrial, commercial, and colonial dominance and later brought on the American challenge in heavy industry.

The collapse of the rubber market and the increasing use of mineral rather than organic matter as primary inputs by core industries, together with Ford's failed attempt to break British control of the world rubber market by establishing plantation cultivation of rubber in the Amazon, left Brazil-nut groves and the commercially marginal remains of the rubber trade as the Amazon's only links to world markets from 1920 to 1950.

In 1950, though, the Amazon was incorporated into the scale-driven world markets for mineral raw materials. The geopolitics of the Cold War and the crucial role of steel in the postwar reconstruction of the European and Asian capitalist economies had generated U.S. political and commercial interest in sources of manganese outside of the Soviet Union. Politically and economically powerful Brazilian groups responded to Bethlehem Steel's blandishments and formed a large company, with substantial Brazilian government guarantees of U.S. Export-Import Bank loans, to export manganese from Amapá, the isolated and sparsely populated territory on the northern shore of the Amazon delta and one of the few locations in the world with large deposits of industrial-quality manganese.

Manganese is an essential component of the production of both Bessemer and open-hearth steel. It is therefore critical to both military and productive competition. It occurs in economically viable deposits in only five countries. The largest known deposits at the end of World War II were in Soviet Georgia.

The Cold War made the U.S. government and U.S. firms very anxious to secure deposits relatively close to the United States. Manganese is used in relatively small proportions to coal and iron—14 pounds per ton of steel—and is priced high enough that transport from the most remote parts of the globe is commercially viable. The Serra do Navio deposits were highly concentrated in an area twenty miles from a potential deep-water port. Construction and then operation of mine, railroad, and port generated considerable prosperity within this coastal enclave and huge profits for the concentrated capital that controlled it, but these effects were spatially and materially more limited than the effects of the rubber boom had been.

Between 1920 and 1950, transport technology and economies of scale advanced rapidly, as did the world's appetite for steel products and the scale of the financial institutions that aggregated finance sufficient to the economies of scale driving this exponentially expanding cycle. These processes finally made it possible for the core nations to incorporate the Amazon as a supplier of the extremely low-value/high-volume goods extracted from Carajás.

Carajás shows how technological innovations generate political and financial forces that further advance the materio-spatial intensity and extent of the world-system. The twenty years of complex political and economic negotiation and struggle that preceded its 1985 inauguration reflected the complex interactions of finance and politics triggered by the volume and scope of matter and space involved. As the largest iron mine in the world, Carajás merged two parallel trajectories—the global trajectory of the technologically driven, scale-augmenting, sequentially cumulative cycles of material intensification and spatial expansion in the world economy and the local trajectory of the demand-driven local extractive economies of the Amazon. The Companhia do Vale do Rio Doce (CVRD), the public-private Brazilian firm that already exported more iron ore than any other company in the world, now owned and operated the world's first globally competitive iron mine.

The Carajás project was simultaneously catalyzed by and catalyzed competition for trade dominance among the United States, Japan, and the European Union. CVRD, the Brazilian national state, and the state of Pará all sought their different interests in the implementation of these projects, as did steel firms and financial institutions in the various competing nations. The competition and collaboration between these various firms and states engendered economies of scale with expanded technologies and infrastructure. These economies of scale devastated local environmental, financial, productive, and political systems. The resistance and resilience of these systems had already been weakened by the legacy of the rubber boom and other extractive cycles.

The technological innovations and economies of scale in mining, transport, and production that made Carajás iron ore competitive in Japan constituted a major component of the technologies driving the late twentieth-century phases of globalization. The design, finance, construction, and operation of the Carajás mine represented the latest step in the globalization of iron and coal markets by technical advances in the scale of mining, transport, and processing. Along with oil, coal and iron are now the most voluminously consumed materials in industrial production. Competitive industrial performance therefore depends on their being relatively cheap, and the nation that receives them most inexpensively thereby gains a huge competitive advantage. Core firms and states are therefore highly motivated to invest considerable effort and capital to develop technologies, organizations, and geopolitical capacities that help to reduce these costs.

Over the three decades before the first trainload of iron left Carajás in 1985, Japanese firms—in collaboration with state agencies, sectoral associations, and national banks there and with Korean firms similarly supported by Korean state agencies and banks—had competed with European firms and state agencies tied to the European common market for steel to increase their shares in world shipping and world iron and steel markets. These competitive strategies reiterated those of earlier aspirants to trade dominance, but they were also molded by a new variant on traditional geopolitics. The United States promoted finance for these efforts—directly through its own Export-Import Bank and through the Marshall Plan, and indirectly through the World Bank and through contracts to supply its wars against North Korea and North Vietnam —as part of a broader campaign to strengthen the capitalist European and Asian steel and shipping industries against Soviet industrial, financial, military, and diplomatic power.

By 1980, Japanese shipyards had increased the maximum cargo sizes for oil tankers and for dry-bulk iron ore carriers fivefold over 1950 levels (see fig. 3.2). These huge increases in size and weight were made possible by Japanese success in mass-producing high tensile steel inexpensively enough to build these ships at competitive cost. The cheap mass production of high tensile steel in turn required continuous smelting and casting in computerized smelters large enough to sustain extremely high temperatures for precise amounts of time. Japanese innovations in the strength, scale, and control of heat in smelting and in construction created and reinforced synergies between steel production and shipbuilding. State banks and ministries subsidized loans to both industries and invested in huge new ports in enormous industrial parks on lands the state, financial institutions, and steel companies collaborated in building out into deep harbors.

These same firms made long-term contracts and these same banks made foreign loans to encourage selected (increasingly distant) potential suppliers of iron and of coal, first in Australia, then in Brazil, and finally in Canada, to build the ports and railroads needed to load such enormous vessels with iron and coal. These huge scale increases in transport technology and infrastructure, together with the exporting nations' assumption of debt to build that infrastructure in their own territory, allowed Japanese steel companies to import raw materials from halfway around the world more inexpensively than U.S. Steel could ship iron ore from the northern side of the Great Lakes to the southern side.

Peripheral state willingness to assume part or all of the cost of building the infrastructure adequate to support the rapidly expanding scale of raw materials exports cheapened and facilitated globalizing processes that directly benefited the core while deeply prejudicing peripheral raw materials exporters. Raw materials exporters assumed the costs of investments that would ultimately weaken their bargaining position and reduce the rents they could collect from extraction while vastly increasing their foreign debt. This extraordinary result was achieved through an evolving hegemonic discourse that invoked the nineteenth-century experience of the United States to claim that natural resource extraction would lead to economic and political development. The U.S. firms and state had initiated this discourse after World War II to encourage colonial countries to seek independence from Britain and France and then to allow U.S. foreign direct investment (FDI) in minerals (Bunker and O'Hearn 1992; O'Hearn 2002). The Japanese firms and state appropriated and modified the same discourse to convince peripheral states and firms that national equity control of extractive projects would bring them the same wealth and profit that monopoly control had earned for U.S. transnational firms. Well-funded private and public Japanese consultant firms specialized in the production of technically complex and glossily published economic studies that invoked mainstream economic theorists and used mathematically sophisticated models to project the extraordinary increases in regional prosperity to be gained by the mines, railroads, hydroelectric dams, ports, and smelters that they were proposing.

The promulgation of this discourse was extraordinarily effective in providing core firms and states with a cheap and steady supply of raw materials and in correspondingly reducing the rents and prices their peripheral suppliers received. During the three decades that the Japanese were developing the technology and finance that made bulk transport cheap and efficient enough to support the globalization of iron and coal markets, the economic policies and aspirations of numerous peripheral nations, including major exporters of minerals, were being realigned in attempts to emulate the economic and political practices of core nations.

Even before World War II, firms and the state in the United States realized that they could procure such supplies more easily and cheaply if they could persuade peripheral states and firms to organize the extraction and transport themselves (Staley 1937; Bunker and O'Hearn 1992). The globalization of iron ore markets emerged from complex processes of collaboration and competition between firms and states in the core as they first induced and then responded to local actions and initiatives in the periphery to profit from their needs for raw materials. Competing core states strove to influence economic and political policies of resource-rich nations in ways that allowed them to adapt to and exploit the materio-spatial characteristics of the new extractive peripheries at competitively low costs. They then had to adapt to the unintended long-term consequences of their own actions, as the desires of peripheral nations for autonomous industrial development, capital accumulation, and improved living and political standards encouraged them to take their own initiatives and mount their own resistance to their growing perception of inequitably uneven development and distribution of wealth.

The United States, for example, after World War II had advocated the national independence of colonies and the industrialization of less-developed nations in order to foster managerial autonomy and open trade there. This strategy emerged directly from recommendations by the Council on Foreign Relations aimed at lowering the costs of access to peripheral raw materials (Bunker and O'Hearn 1992). These U.S. policies helped create incentives that eventually generated fiscal incentives and tariff protections for Import Substitution Industrialization (ISI). ISI principles contradicted the FDI preferences and free-trade principles of the U.S. government and of U.S.-based multinational firms. By the 1960s, this contradiction led to broad resource nationalism in peripheral nations, many of which nationalized their foreign-owned mines and processors; more demanded renegotiation of their FDI contracts.

These changes eventually enabled the peripheral states to claim greater control over extractive exports. Their claims threatened to raise the costs and reduce the security of access to raw materials until core firms and states devised new political and economic strategies to lower costs and increase stability again. These trends deterred major mining companies from investing in exploration or new projects.

Together with the multilateral banks established by the victorious nations after World War II to maintain sufficient liquidity of capital to ensure sustained international trade, core states and their international banks—particularly the Export-Import Banks of the United States and Japan—reacted quickly and decisively. They mobilized new forms of credit, supported by joint ventures and long-term purchase contracts, for peripheral raw materials suppliers. These loans and joint ventures encouraged local investment to replace the FDI capital that foreign firms were no longer willing to risk and that peripheral states were increasingly reluctant to accept.

These instruments, collectively called New Forms of Investment (NFI; Oman 1984, 1989), displaced risk and debt for scale increases in extraction and transport from core consumers to peripheral suppliers. This cheapened the cost of raw materials in the core and effectively lowered the ground rent paid for raw materials extracted in the periphery. Firms and states in the core thus provoked economic and political changes in the periphery to cheapen their access to raw materials. The World Bank was implementing the logic of these new forms of international investment and credit when, responding to core-wide concerns about potential raw materials shortages, it mobilized its considerable technical, financial, and diplomatic resources to coordinate and supplement loans to CVRD from Japan, Germany, the European Union, and Korea to bring the Carajás mine into operation. (The timing of the World Bank's intervention and participation in the loan negotiations in 1982 and its willingness to relax its procedures to link disbursement of the loan to the completion of sequentially specified stages suggest that the World Bank efforts in this case may also have aimed at making Brazilian default on loans less likely.)

The NFI strategies that shaped the loans and long-term contracts emerged from core state, firm, and bank responses to a wave of resource nationalism in the resource-rich peripheral nations that supplied them with essential raw materials. Because the NFI strategies appeared to accommodate demands of peripheral nations to share in the control of extractive projects within their territories at the same time that it reduced capital risk of core firms, NFI was particularly attractive to core nations that were striving to expand their shares of world trade against U.S. domination. In that sense, such strategies represent a core rival taking advantage of peripheral initiatives, and they suggest that competition within the core motivates a search for ways of attracting peripheral nations to share in the extractive enterprise. The aggressive promulgation of NFI strategies to multiple potential mineral suppliers accelerates globalization of the sources of raw material. At the same time, it reduces the cost to the ascendant and the dominant core economy by distributing the infrastructural costs of and reducing the effective rents paid for the new extractive projects that provide them with the raw materials they need to continue expanding their production.

The unprecedented size of this loan (US$ 3.7 billion) was proportional to the unprecedented scale of the mine and of the processing and transport infrastructure it required to export its iron to global markets. The scale of this infrastructure emerged from (and was justified by) the economies of scale in transport that had brought about (and were made possible by) the latest stage in the material expansion and spatial extension to a global and ocean-based, rather than regional and land-based, market for coal and iron. The technologies that supported these economies of scale emerged from state-sector-firm collaborations in Japan, Korea, and Europe aimed at increasing their respective national shares of the U.S.-dominated global markets (Bunker and Ciccantell 2003a, b). Japanese firms and the Japanese state proved much more agile than U.S. firms and the U.S. state in their responses to the emerging resource nationalism of peripheral nations. They actively encouraged local control over the projects that would supply them with raw materials, and they provided long-term contracts that their potential suppliers could show to the banks they were asking for finance. Japanese encouragement of CVRD's commitment to build the railroad all the way to the deep-water port at São Luís and the resulting dissolution of CVRD's partnership with U.S. Steel were both typical of the kinds of strategy that the Japanese used to cheapen their own access to raw materials—and threaten that of their competitors.

Within the Amazon, the proposal to construct this infrastructure had generated intense opposition. The Tocantins River flowed 150 kilometers to the east of the Carajás deposits, through groves of Brazil-nut trees whose products were shipped down the river for export from the port of Belém, the capital of the state of Pará. Original plans for exporting the iron through a partnership of U.S. Steel and CVRD had specified a rail line from the mine to the river, then barge transport to Belém where the ore would be loaded onto oceangoing ships. This sequence of transport would have limited ships to a size small enough to navigate the Great Lakes, where U.S. Steel still had its smelters.

U.S. Steel strongly opposed CVRD's preference for the much more costly project, the funding of which the World Bank eventually coordinated. Instead of going out through the river port of Belém, the iron would be exported from an ocean port in Maranhão. Belém's port could accommodate at most 60,000-dwt (deadweight ton) boats; Maranhão could potentially handle over 400,000-dwt boats. Exports to U.S. Steel in the United States would not travel distances long enough for the economy of scale in ship size to offset the additional fixed and demurrage costs of the larger ports.* European and Japanese ports, however, were sufficiently distant for this to be the case.

World Bank coordination of multiple consuming-country loans finally provided CVRD access to global markets. Multilateral financing for Carajás to a large firm partially owned by a peripheral state, secured by long-term contracts to sell ore to Japanese, Korean, and European competitors, was the largest and most significant of the new global financial instruments that were displacing the system of FDI in mining and transport infrastructure that the U.S. had used to achieve its midcentury control of world steel. Subsidies and fiscal incentives from both the national and the local state supported CVRD's ability to guarantee and administer this loan. This financial guarantee effectively devolved to the Brazilian economy the costs of Japanese, Korean, and European strategies to challenge U.S. dominance. In addition to this direct reduction of tax revenues, the loan enabled CVRD to build a railroad and a port, which deprived the State of Pará of the direct taxes and the indirect income that export from Belém would have provided.

State and firm in Brazil were both key actors in the resulting financial and managerial processes. Interview and archival research in Belém, Rio de Janeiro, Tokyo, and Washington, D.C., on the origins of this controversy strongly suggests that the ambitions of CVRD and of various agencies of the Brazilian national state to preserve and expand Brazil's share of world iron ore markets had made them very open to ideological and contractual inducements from Japanese and, to a lesser extent, European interests to support the coastal rather than the river option for export.

At the same time, engineers and economists in CVRD and in these same state agencies, largely trained in the United States and in Europe, took very seriously the conventional wisdom that attributed European and American success in industrialization to their nineteenth-century iron-, coal-, and steam-based economies. They had little social or economic reason to dwell on the

---

*The materio-spatial regularity in this case is that the water resistance against the hull increases as the square of the depth of the hull, while the volume of cargo increases as the cube of the depth of the hull. Fuel requirements per unit of cargo thus reduce with increased size of ship. At the same time, labor requirements do not increase at the same rate that ship size does. These two constants create consistent economies of scale in shipping. These economies of scale are offset, however, by the much higher sunk capital costs of the infrastructure required for docking and loading cargo and for the greater costs of in-port delays incurred by a larger, more costly ship. The economies of scale are thus only realized where the voyage itself is long enough to offset the extra fixed costs of in-port handling.

possibility that the experience of a century earlier might have changed with the huge reduction in transport scale and cost—neither, certainly, did their Japanese, European, and American counterparts with whom they were negotiating. Such a generalized consensus among powerfully situated actors is a frequent—perhaps essential—component of a hegemonic discourse that persuades all participants that the ends that favor the most powerful among them will in fact benefit all. Thus Brazilian agents of the federal state and of CVRD effectively silenced or marginalized protests within the state of Pará against the scale of the proposed mine and transport complex and its devastating local fiscal, economic, social, and political impacts. They cited leading national economists who claimed as general principle the nineteenth-century U.S. and European experience of transport and extraction promoting agriculture and industry. They also made threats ranging from lawsuits against the state government to the withdrawal of fiscal transfers from the national level.

CVRD and Brazilian federal agencies, with World Bank support, thus played into and exploited Japan's strategies to dominate the world's steel and shipping markets. It is very likely that, without active CVRD initiatives, the incorporation of Carajás into a globalizing market for iron ore utilizing large-scale dry-bulk shipping would not have occurred and that the trajectory and rate of globalization would have been at least different, and possibly reduced. CVRD's collaboration was essential to Japanese, Korean, and European moves that globalized these markets in order to eclipse the U.S. dominance of the hemispheric steel and transport sectors. In other words, peripheral participation in this case is an essential component of globalization, but that participation is catalyzed, encouraged, and shaped by competing core nations.

The scale of technologies for mining, transporting, and smelting, which had responded to and been fueled by the topography and geology of the Great Lakes region a century earlier, had by now surpassed the capacities of these same materio-spatial features. The huge boats that the São Luís port was deep enough to handle and the enormous machines that the Carajás deposits were big enough to support had moved the world's steel industry well beyond the continental economies of scale that the Great Lakes iron and steel complexes spawned. The industry's quest for cheap raw materials had by now progressed from regional river-based networks of eighteenth- and nineteenth-century Europe to lake-based boat and land-based rail systems of nineteenth- and early twentieth-century America to the ocean-based hemispheric and finally global supply systems that arose at the end of the twentieth century.

CVRD's actions in these undertakings demonstrate three interdependent conclusions. First, social, political, and economic struggles and strategies in materially and spatially differentiated peripheries structure differentially the processes and outcomes of inter-core competition and collaboration. Second, preferences and decisions of peripheral actors have consequential roles in restructuring peripheral-core relations.* Third, increasing economies of scale in the extraction, transport, and processing of the most voluminously consumed raw materials catalyze the episodic but cumulative spatial expansion and material intensification of the world-system through a succession of distinct systemic cycles of accumulation (cf. Arrighi 1994).[†]

The materio-spatial processes and relations that thus structured political, economic, and financial relations around the construction of the Carajás plant and transport infrastructure coincided with massive scale expansions in the steel industries of several nations. These included (1) the peak of Japan's expanded production of steel and of ships to dominate world trade in both sectors, (2) Korea's entry into the world steel market through the construction of the largest integrated steel plant in the world, and (3) the relocation of European steel processing plants to coastal locations more suitable for the larger scale of smelting and transporting needed to compete with the drastic cost and price reductions of steel achieved by Japanese and Korean companies (Jörnmark 1993).

These scale increases in core smelting technologies (and the resultant cheapening of steel prices and increase in world consumption) combined with major politically driven changes in global finance that enhanced market opportunities and pressures for increased iron and coal extraction. OPEC's success in raising the oil rents of its members flooded international banks with petrodollars, and banks anxious to keep their deposits circulating and earning interest prompted a flood of easy credit for large-scale foreign revenue-earning projects such as metal mining and processing in the periphery.

The Plaza Accords, orchestrated by the U.S. Commerce, Treasury, and State

---

*Note that CVRD's preferences affected not only Brazil's position in the world-system but also the entire world market for iron and steel. The addition of Carajás increased world supplies to a highly inelastic market, with implications for all exporters of iron and for all industrial processors of steel.

[†]Japanese participation in this multilaterally shared loan was itself a new step in the strategies to reduce capital exposure and risk through joint ventures and long-term contracts that the Japanese had perfected in their dealings with Australia. In this case, the Japanese succeeded in extending the oversupply from excess capacity in world mines by responding to and participating in the core-wide preoccupation with resource scarcity (Bosson and Varon 1977; Bunker and Ciccantell 2003b).

Departments a few years after the two oil price shocks to ease American balance of payment problems, which were exacerbated by President Ronald Reagan's fiscal policies, greatly increased the strength of the Japanese yen against the U.S. dollar. Firms and states that had borrowed yen to finance projects that sold products into dollar-based markets suddenly held greater debt and less revenue. Accelerated U.S. international debt, also a product of Reagan's policies, raised interest rates abruptly, exacerbating the increasingly high cost of peripheral nations' debt and decreasing their willingness to incur further debt in large-scale extractive export projects.

In summary, the scale expansion and relocation to coastal sites that the Japanese, European, and Korean steel and ship-building industries undertook directly drove the spatial expansion and material intensification of the raw materials markets; OPEC nations concentrated and then dispersed a large portion of the world economy's liquid capital; Japanese and European strategies to devolve the infrastructural costs of the expanded scale of transport technology were enhanced and facilitated by the resulting easy credit; and the Plaza Accords enormously raised the exchange rates of the Japanese yen, increased the interest and capital costs of all loans made in yen, and lowered the real value of international sales of iron, which were transacted in dollars. These financial and political processes were essential to the realization of the new, expanded technological scale and to the restructured extraction, transport, and commerce of raw materials that they required, but in both instances the finance and politics were secondary to the primary imperative driving these processes: the procurement of technologically necessitated raw materials of a certain quality and volume at competitively low prices.

The vast volume of petrodollars that originated much of the global fiscal crisis of the 1970s and 1980s was itself founded on the absolute dependence of core industry on peripheral sources of petroleum. At a different level of market structure, but also highly relevant to the nature-society dynamic we are considering here, the excess accumulation of liquid capital in Japan, which created the fiscal crisis that the Plaza Accords were designed to ameliorate (Murphy 1996), was itself initiated by Japanese success in reorganizing the spatio-material distribution and social organization of the world's metals markets in ways that fostered Japanese dominance of world steel and merchant marine markets. The cumulative result of these interdependent financial, spatial, and material transformations was the enhanced indebtedness and subordination of raw materials–exporting peripheries to raw materials–importing nations of the core (Bunker and Ciccantell 2003a, b).

This subordination, however, involved the active participation of local agents in the creation of the capital-intensive infrastructure as well as the adoption of new technologies with massive economies of scale in mining and transport that supported the core advances in the scale of smelting and refining. CVRD decided to respond to the opportunities these new technologies opened in the Japanese market and to the facilitated access to huge credits opened by the concurrence of the world debt crisis and core country worries about future raw materials shortages by contracting huge loans to (1) increase the extractive capacity of the mine, (2) build the high-capacity rail line to the coast, (3) develop a port adequate to take ships so large that only Japanese ports and the port of Rotterdam could unload them, and (4) engage in joint ventures with Japanese firms in ore carriers of over 450,000 dwt.

The Effects of Iron, Brazil-Nut, Aluminum, and Electricity Extraction in Amazonia

The development and manipulation of a hegemonic discourse to promote extractive enterprises that benefit the core while prejudicing the peripheral areas affected is even more dramatically illustrated by the effects of the aluminum industry on some of the areas affected by the iron mine. At the mine site itself and along the rail line, the iron deposits' materio-spatial features intersected with the capital cost required for the scale of plant and transport infrastructure to shape CVRD's economic and political decisions. This intersection of natural, social, and financial forces constrained CVRD to coordinate the construction of the operation's constituent parts so that they would start as close to simultaneously as possible. No component of the system could function until all were in place, so temporal disjunction in their completion would increase interest costs on what was in any event an enormous investment. The combined material, temporal, spatial, and financial pressures led to a highly concentrated use of labor in a relatively restricted space over a precisely limited period of time.

CVRD's intensive labor needs prompted both formally managed and spontaneous migration of labor to Marabá, the *município* where Carajás was located. The materio-spatial consequences of two other extractive economies, Brazil nuts (already established and particularly local in its social and political relations) and the generation of hydroelectricity to smelt aluminum (novel and global in both market and organizational style) exacerbated the intense conflicts over space and matter there. In this section, we will consider what the political strategies and struggles of the diverse groups and agencies involved in these conflicts reveal about processes of globalization.

Until 1980, Marabá was the major source of Brazil nuts in the Amazon and therefore in the world. Unlike rubber trees, Brazil-nut trees grow in dense groves. In other respects, though, the spatio-material features of the *castanheira*, or Brazil-nut tree, favored the adoption of the basic material, financial, and social relations of the *aviamento* system. As in rubber, the trade was organized into large *aforamentos*, or long-term leases from the state, held in such a way as to allow potentially violent control over access to and transport out of the groves. As the rubber merchants had, the Brazil-nut merchants used this system of property and labor relations in order to secure the return and profit on the capital advanced to transport and sustain labor during the months of the harvest.

Migrants who failed to find employment in mine and rail construction and, later, employees who were dismissed when all of the huge construction projects finished at the same time could instead occupy and exploit the lands in and around the Brazil-nut groves. The trees themselves depend biologically on the clustering of other trees, both for protection against the wind and to support a bee essential to their pollination. Thus, any agricultural use of land in or near a grove can drastically reduce the production of Brazil nuts. The owners of the long-term leases publicly announced that they would respond to any challenge to their control of property and of labor with violence.

With direct encouragement and technical assistance from Japanese consulting firms contracted by the Japanese state, the Brazilian federal state set out to support CVRD's search for international financing for Carajás. In 1980, it established a fiscal incentive program to attract businesses that would share some of the infrastructural costs of the mine's transport system and solidify the theoretical claims that extraction would generate agricultural and industrial development in the area.

Violent conflict over land and resources threatened to disrupt the smooth implantation of the mine and of the other enterprises that CVRD and the national state were hoping to promote around it. The state responded to this threat—again, under Japanese urging—by establishing GETAT, a land-policing quasi-military organization that was given extraordinary powers and authority. GETAT could review the legal and administrative procedures of Pará's *aforamentos* and thus, even though it formally recognized the local state's ownership of those Brazil-nut groves it deemed correctly and legally leased, GETAT preempted local state capacity and authority in Marabá.

The federal state thus used its economic, political, and military powers to bring about the effects that the Japanese were claiming would follow naturally, thus fulfilling the predictions used to legitimate the huge expenditures in a project to export local mineral wealth at the lowest possible cost. These federal decisions were implemented at direct cost—both political and fiscal—to the local state and indirectly to the national population as a whole. CVRD and the other capitalist firms attracted into the region to use CVRD's new infrastructure also benefited directly. Less directly, but more fundamentally, the nations of the global core, particularly Japan, benefited from the consequent lowering of the prices of their iron ore imports.

As in the rubber economy, the materio-spatial characteristics of the Brazil-nut economy directly influenced political and legal institutions and processes. The initiation of the iron-mining economy, however, brought political, economic, and legal processes that were fundamentally at odds with those of the Brazil-nut economy into the same space. The Brazilian national state, under intense pressure from CVRD and from Japanese financial and planning agencies, intervened to reduce the autonomy of the local state and of the *município* of Marabá and to restrict the political power of the old Brazil-nut oligarchy.

The uneven conflict between Brazil-nut and iron economies was soon complicated by Japanese searches for two other raw materials—aluminum and the hydroelectric power to generate sufficient quantities of electricity to smelt it. The material characteristics of aluminum brought entirely new ecological and economic pressures and imperatives to bear on the region—and on the other sources of aluminum and electricity around the world.

Coordinated by the Keidanren, Japan's powerful national business association, Japanese aluminum firms and sectoral associations, banks, and government agencies were attempting to globalize aluminum smelting. They used NFI techniques to relocate this most capital- and energy-intensive phase of aluminum processing from Japan to peripheral nations with rich sources of bauxite and the appropriate topography and hydrology to generate electricity. They presented highly polished projects that showed how aluminum-exporting nations could share in the surplus profits that the big six transnational aluminum corporations had been enjoying since World War II. They offered to join in joint ventures with peripheral states willing to invest in dams that would provide cheap electricity to smelters under long-term contracts (Bunker and O'Hearn 1992; Bunker 1994a; Bunker and Ciccantell 1994).

CVRD accepted an invitation to form a joint venture with seven Japanese aluminum companies in Alunorte, an alumina refinery, and Albras, an aluminum smelter, downriver from Marabá. The Brazilian government and the public electric company, Eletronorte, were induced to build a huge hydro-electric dam across the Tocantins River to supply the aluminum smelters with cheap electricity. The Japanese government offered at first to participate in the financing of this dam, but after the Brazilian government had publicly and fiscally committed to the project, the Japanese withdrew their offer, alleging that the project they had themselves proposed was not economically viable. Eletronorte assumed huge debt to finish the project and guaranteed long-term prices for the electricity transmitted from the dam to the smelter. These prices turned out to be well below the cost of production and interest charges. The resulting drain on Eletronorte's revenues impeded the company's projects to extend transmission lines to other regions and other users in the Amazon and even to maintain its existing services. Federal pressures to favor CVRD and its international clients pushed the public company organized to supply power to the Amazon region to the brink of bankruptcy.

Rising energy costs in Japan had induced the Japanese state and Japanese firms to collaborate in moving the energy-intensive aluminum industry out of Japan. Their preferred means of doing this was to create joint ventures with the national states that controlled large sources of bauxite, in whose territories it would be possible to generate large quantities of hydroelectricity. Because bauxite forms through the percolation downward of aluminum and the dissipation of the water-solvent silica that bonds to it, the richest deposits occur in flat areas in the humid tropics. Hydroelectric dams in narrow steep valleys provide the cheapest sources of the huge amounts of electricity required to separate the high valence bonding of oxygen in aluminum (AlO_3).

Locating an aluminum smelter near a bauxite mine therefore makes little economic or ecological sense. The hydroelectric dams that can generate enough energy to smelt aluminum in the flat topographies that produce the best sources of bauxite require huge amounts of water spread over huge areas to compensate for the topographic limits on the dam's height.* Dams in steep valleys can generate electricity more cheaply and with less environmental damage, but such valleys do not occur near large high-grade bauxite deposits.

Japanese NFI strategies to devolve much of the cost of aluminum smelting onto their supplier nations manipulated a hegemonic discourse that rested on

---

*The amount of energy generated is a direct function of the volume of water times the height of the dam ($Kw = vh$).

the same historical arguments that resource extraction promoted linked industries, embellished with the powerful symbolic association of modernity with electricity. The Japanese sponsored and conducted regional economic analyses predicting that providing cheap electricity for smelting aluminum would add value to the bauxite they would otherwise have exported in a raw state—while stimulating other industries that would be attracted by cheap and abundant electricity. The Japanese economic technicians particularly mentioned industries that transformed aluminum into end products as susceptible to such enticements.

This discourse was particularly convincing because of the persistently high price of aluminum and the secrecy with which Japanese firms and state conducted negotiations with multiple nations—including Venezuela, Brazil, Canada, and Indonesia—whose negotiators each believed that their country would be Japan's primary supplier of aluminum. When all the separately negotiated hydroelectric dams and smelters came on line in the early 1980s, excess supply severely depressed world markets. Obviously, the lower prices favored the Japanese firms that fabricated aluminum and left their supplier nations holding massive debt and a greatly reduced capacity to pay the rapidly increasing interest to which they had agreed (Bunker 1994a).

Agencies of the Japanese state—particularly MITI (the Ministry for International Trade and Industry), JICA (Japan International Cooperation Agency), MMAJ (Metal Mining Agency of Japan), and the Export-Import Bank—together with NAAC (Nippon Amazon Aluminum Corporation), the consortium of seven Japanese aluminum firms that had formed to invest in Albras, and the Keidanren, the powerful and prestigious national business association of Japan, collaborated in a sophisticated manipulation of Brazilian aspirations for industrialization of the Amazon to gain local political and financial commitment to build dams. They contracted the International Development Corporation of Japan (IDCJ), a reputable and well-staffed development consultancy in Tokyo, to carry out and publish an analysis of the economic and social benefits of electric generation in the region. These same studies also described ways that the dams could facilitate commercial transport by stabilizing water levels on their rivers. They included elaborate projections of agricultural and industrial growth that these enhanced transport routes would stimulate. They used complex econometric formulas to project the investment and resulting regional income they claimed a hydroelectric dam constructed for a regional vertical integration of all three phases of aluminum production—

mining, refining, and smelting—would catalyze. Finally, they emphasized the economic and social benefits of massive cheap electricity for the region as a whole.

Japanese state agencies, banks, and aluminum firms thus contrived to reduce the cost to themselves of building an economically inefficient dam by inducing Brazil to assume most of the equity costs of these undertakings (Bunker 1994a, b). Eletronorte was able to provide the transmission lines to the smelter, but it could not afford to extend the lines to established communities. No locks were built around the dams, so they now impede the local transport they were originally promised to enhance.

At the time, the aluminum industry was controlled by six huge firms based in North America and Europe that used their oligopolistic power to keep prices high by keeping supply well below demand. The history of these high prices was a major factor in the state's decision to assume the high capital costs of building the dam. CVRD officers who had been engaged in negotiations over the division of costs for the dam and the division of shares, liabilities, obligations, and product of the smelter later complained that the Japanese negotiators had constantly shifted their ground, changing the terms they proposed or were willing to accept (Bunker interview, 1991). The Brazilian negotiators did not realize at the time that the Japanese state, business groups, and aluminum companies were negotiating similar joint ventures with six other nations at the same time and presumably playing the terms for each off against the others. They did not foresee that, when all six of these multiple joint-venture smelters came on line at about the same time, they would vastly oversupply the market. The Japanese thus precipitated the price crash of 1982.

Because the Japanese firms had reduced their equity shares to less than 50 percent in all cases, and because their main profits were from aluminum product fabrication rather than from the sale of primary aluminum, they benefited from the lower costs of raw material, while the states and firms in the aluminum-exporting countries confronted increased debt and reduced revenues (Bunker 1994a). By strengthening the yen, the Plaza Accords hugely increased the costs of the loans that firms and states had accepted to finance the hydroelectric dams and smelters.

The Japanese strategy essentially involved an inversion of the materio-spatial logic of the aluminum oligopoly, which had located smelting, the most expensive and capital-intensive phase of processing, in the industrialized countries. The Japanese strategy was to promote joint-venture smelters in nations where bauxite was abundant and whose states or state-owned companies might be induced to generate and sell electricity cheaply.

By manipulating the resource nationalism and desire for industrial development of bauxite-rich nations, the Japanese relocated the capital-intensive —and energy-intensive—phases of the industry to the periphery at the cost of the peripheral nations. The Japanese promulgated an oversimplified and wildly optimistic version of vent-for-surplus notions that capital from raw materials exports could be invested in processing and transformation in such a way as to "capture" the linkage or spread effects of these materials. The Japanese incorporated these ideas into elaborately prepared and widely distributed technical studies and project proposals in which they claimed that investments in natural resource extraction and processing, transport infrastructure, and hydroelectric generation would maximize economic development.

Japanese firms and agencies of the Japanese government combined these studies with joint ventures in plants and offers of participation in infrastructure to induce the states of aluminum-rich nations to invest in dams and smelters (Bunker 1989b, 1994a; Bunker and O'Hearn 1992). The study proposing the construction of a hydroelectric dam to serve the aluminum smelter in the Amazon ran to seven volumes, printed on glossy paper with ample tables, graphs, and photos. National, state, and provincial governments in Indonesia, Venezuela, Australia, Canada, and Brazil were all seduced by Japanese offers to exploit their hydroelectric potential to smelt aluminum. The Japanese found willing collaborators in promoting these ideas, particularly in CVRD and in those ministries of the national state that were primarily concerned with finance and foreign trade. Multiple decisions at the local level were the necessary conditions for the spatial expansion, material intensification, and organizational and financial transformation that globalized the world's aluminum industry.

Japanese initiatives had catalyzed the globalization of iron and coal extraction and export by stimulating states and firms in distant source nations to invest in mines and transport. Their main intent had been to supply cheap raw materials to increasingly centralized, scale-economic smelters in Japan. In aluminum, the tactics were inverted but the consequences were the same. Instead of exporting bauxite, CVRD (with support from Eletronorte and fiscal incentives, subsidies, and tax relief from both the national and local governments in Brazil) now incorporated electricity, which cannot be transported or even transmitted over great distances, into aluminum. It could then export the incorporated electricity to Japan at prices less than the cost of production. The loss from this unequal exchange, however, did not affect CVRD's balance sheet. Rather, the loss was devolved to Eletronorte—and to the public Eletronorte was created to serve.

As in Carajás, the infrastructure for this chemically complex transport system—hydroelectric dam, more than half of the smelter, plus loading docks and ports, as well as associated roads and electric transmission lines—was financed with debt assumed by CVRD and by the Brazilian and Paraense governments. Tax holidays conferred by the federal state meant that Pará had to finance the additional costs of public welfare and law and order as well.

As with Carajás, this global project had continuing high costs for the local state. CVRD and NAAC later took further advantage of the weak bargaining position into which they had forced the local state. They negotiated additional tax concessions from Pará as a condition of completing Alunorte, the alumina refinery that was included in the original plans, and later Albras withheld an additional share of the value-added taxes it owed the state as it challenged the state's right to impose value-added taxes on exported goods.

Again, the local process is not simply a reflex of a global phenomenon but rather directly drives the transformation of the global system. Japanese initiatives globalized production by decentralizing what had been a highly concentrated monopolistic system of smelting to the sites of raw materials supplies and then inducing the host nation to share in the capital costs. Both strategies expanded the commercial arenas of these metals by creating excess capacity and cheap transport. Excess capacity led to excess supply and thus lowered prices. Cheap transport increased the number of potential sources, thereby creating competition between suppliers who earlier would have monopolized spatially separate markets. This competition enabled the importers of raw materials to bargain down the rents they paid for these materials. Thus the supplying nations were induced to invest in outcomes that reduced the prices and the rents they might receive from the raw materials extracted from their ground.

Japanese firms, sectoral associations, financial institutions, and state agencies thus simultaneously transformed and globalized markets (Bunker 1992, 1994a; Bunker and O'Hearn 1992; Bunker and Ciccantell 1994, 1995a, b, 2003a, b; Ciccantell and Bunker 1997, 1999, 2002) for iron, aluminum, and electricity, all commodities whose markets had been regional or national. In the process, they also catalyzed the expanded scale of extraction and processing, both in Brazil and globally.

To succeed, Japanese access strategies had remained highly responsive to a wide range of natural and social local dynamics—from the political, cultural, and ideological to the physical, chemical, and topographical.* They simultaneously had to take into account (1) the physical, chemical, and spatial features of the material itself; (2) the patterns of supply, demand, price, and spatial organization in the world market; (3) the political organization, political culture, and economic ideologies of the exporting nations; (4) the market as structured by earlier dominant economies; and (5) the responses of the exporting nations to the specific measures the Japanese were taking to restructure the market to their own advantage (Bunker 1994a).

By tracing changes in the scale of extraction and transport of Amazonian extractive economies, as well as changes in their finance, ownership, labor organization, and political institutions through progressive reduction of value-to-volume ratios across five hundred years of world-system history, we have seen that the dependence of core economies on expanded volumes of matter from the Amazon increased, while the dynamics that expanded these volumes progressively undermined the Amazon's social and ecological integrity and worsened its terms of trade. These effects progressively reduced its status and prosperity relative to the nations that imported its natural products. The Amazon moved from one environmentally destructive, demographically dislocating, and ultimately self-limiting extractive economy to another. In contrast, the nations that imported its raw materials increased their production, their population, and their wealth by devising new technologies; by generating the new political, financial, and scientific institutions those technologies needed; and by using the power of these new institutions to procure greater volumes of raw material more cheaply and more predictably.

We showed through this history how three levels of reality—abstract physical laws, regular material processes, and social history—molded the interactions of society and nature that incorporated the Amazon into the world economy. We now broaden our focus to examine more generally how economic, ecological, and historical processes interacting across these three levels create and expand the global inequalities between extractive and productive economies.

Social production depends absolutely on natural production—even as it

*Japanese negotiating strategies varied in Australia and in Brazil in ways that simultaneously took into account the very different relations of central and local state power in the two countries and the differences in distance between iron mines and the ocean (Bunker and Ciccantell 2003b).

subordinates and destroys it. This contradictory relation of society to nature drives the subordination and immiseration of the extractive economies that mediate between social and natural production. This raises a critical question: why do human communities in resource-rich ecosystems continue to engage in trade that is demonstrably prejudicial to their societies and to the environments that sustain them?

To answer that question, we examine how agents of core states and firms manipulate theories and promulgate ideology, especially economic but also cultural and political, that legitimate extraction by claiming it will generate industrial development. These claims, which form the basis of a globally hegemonic discourse, mystify the differences between extraction and production in ways that encourage the acquiescence of powerful peripheral groups—both economic and political—to the growing ecological and economic imbalances created by the expanding flow of natural products from many ecological systems into a few industrial systems. Agents and intellectuals of core states, firms, and universities theorize highly contingent third-realm narratives of their own nations' successes in transforming extraction into production as if these constituted abstractly universal first- and second-realm laws of economic development. Agents and intellectuals of peripheral states and firms are complicit in the transformation of these mystifying ideologies into hegemonic discourses.

The Amazon's Extractive Cycles in Comparative Historical Perspective

The strategies of earlier ascendant nations—the Portuguese, the Dutch, the British, and the Americans—for securing cheap and steady access to raw materials configured these complex determinacies in varied combinations and interactions, but they are consistently and regularly observable in these different configurations. The history of the Amazon shows that materio-spatial attributes of raw materials and of the location from which they are extracted and exported determine both the economic and political organization of the societies formed around these locations and the technologies of transport and production that transform these raw materials into useful and saleable commodities. These determinacies characterize most extractive economies (Albion 1926; Innis 1956; Bunker 1985, 1992; Barham, Bunker, and O'Hearn 1994; Bunker and Ciccantell 1995a, b, 2003a, b; Ciccantell and Bunker 1999).

Histories of technology and economic development confirm that the physical features of raw materials similarly constrain the technical and social processes of their processing and combination at all points along the chain of production (Marx 1894; Wittfogel 1929; Mokyr 1990; Misa 1995; Landes 1969; Rosenberg and Birdzell 1986). These physical features are created at the site of natural production, which is also necessarily the site of their extraction. In the course of industrial capitalism, however, the invention, engineering, and capitalization of these technologies have increasingly been located in the regions that import and transform raw materials rather than in the regions that extract and export them.

Control of these techniques and of the social-organizational, political, and financial systems that implement them therefore tends to remain in the regions where these materials are transformed and consumed. So does the major share of the profits that accrue from these processes. Scale economies in the core promote the expansion of production, profits, state revenues, and social welfare in the nations competing for trade dominance, while the diseconomies of scale in extraction generate debt, impoverish and weaken the source state, concentrate wealth and income, and eventually deplete the resources on which they are based. This tendency has accelerated as the space across which raw materials are transported has grown to global proportions.

By problematizing the origin and dynamic of this paradoxical disjuncture, we aim to identify a coherent set of mechanisms that explain both the historic growth in the disparity between extractive and industrial economies and the ways that this disparity contributes to (and results from) the mechanisms that drive globalization. The history of the Amazon illustrates the determinacy of space and matter in specific locations in the periphery. Our next task is to extend these lessons from a raw materials–exporting periphery to an analysis of the sources of political and economic power that can explain how the trade-dominant nations of the core manage to control and exploit the multiple sources of the raw materials on which their wealth and profit depend.

Each of the extractive economies we have considered has shown us how local materio-spatial relations and processes in the Amazon intersect with, and partially constitute, the world-system as it transforms and is transformed by systemic changes in cycles of accumulation. The degree of attention we have paid to locally specific materio-spatial attributes, though, has provided us with far more detail and a more complex narrative than we could possibly synthesize into an analysis of the global economy if we gathered similarly complete information about all of its spatially distinct component raw materials–

supplying regions. Nor is there any way to draw a representative sample of so few cases with so many complexly interdependent variables. Instead, we have looked for patterns in the sequence of extractive economies and in the ways that local social organization in the Amazon has responded to both the local material and space and to the opportunities and pressures that the world economy, with its changing markets and technologies, created within that particular configuration of social and political organization, space, and matter. Our task now is to see how we can extend our understanding of this local component of global systems into insights about the systems as wholes.

We can analyze the detailed patterns we perceive in our concrete account of a particular series of events—that is, extraction and export of specific raw materials from a specific place at a specific time—within a more abstract account of how successive nations' competitive strategies for trade dominance led to technological change and increased scale of production. This will allow us to incorporate the richly detailed but necessarily ideographic and complex story specific to directly linked events and processes in a single place to the nomothetic abstractions that allow comparison between sets of indirectly and not so clearly linked events and processes in multiple places.

We will thus be able to base our explanations of the hugely abstract phenomenon of globalization on concrete, materially observable and thus directly comprehensible processes. In other words, by theorizing the historically changing interaction between specific configurations of matter and space on the one hand with specific social organizations on the other as both are incorporated into the world economy, we can generate abstract concepts and mechanisms that will allow us to make sense of how other raw materials—extracting peripheries interact with dominant economies and how dominant economies adapt to the diverse material, spatial, and social formations that they must exploit in order to achieve and maintain their dominant positions.

This is possible because the discrete natural features of specific sites *do* manifest regular underlying processes and mechanisms that can be described abstractly by the rules and laws of physics, chemistry, geology, hydrology, biology, and so on, even though they combine and configure differently in each site. Similarly, the discrete social relations that respond and adapt in order to exploit these natural features also manifest regular underlying processes and mechanisms that can be described in general theories of economics, finance, labor relations, political institutions, and so on, even though they also configure differently in different social situations and organizations.

Natural, or materio-spatial, regularities—immanent from the first and second realms of reality—have broader explanatory power than social ones, which are subject to the far more contingent narratives of the third realm. The physical and chemical properties of iron or aluminum, for example, both generally and in particular deposits, are constant over time. The human uses of these physical and chemical constants vary over time, however, conditioned by historically observable, physically and chemically measurable, changes in technology. These technologies emerge from social intention and organization, which cannot be explained with the same degree of generality and invariability as the heat, pressure, sedimentation, and hydrological dissolution that form iron or aluminum. On the other hand, because they emerge from the interaction of social and material forces, economic processes are more regular than purely social or political processes but less regular than purely physical processes. These different degrees of observable regularity across different domains of material, spatial, social, and financial process distinguish the potential generality and thus the explanatory status of regularities in each domain.

These variations and distinctions do not constitute an impediment to comparative historical analysis; rather, awareness of their different explanatory statuses allows comparative analysis of the relative weight of natural and social factors as their interaction brings about site-specific and global economic and ecological change. Attending to the ways that these regularities configure in different places and at different times will help us to compare local systems across very different scales of economy in production and transport in widely varying spaces and eras. We need to do this kind of comparison in order to understand how globalization has occurred and how it has increased the inequalities between the world's regional economies.

In brief, we will use our richly detailed material knowledge of the Amazon's different extractive economies to ground our understanding of how the global economy has evolved. This will enable us to identify social and physical regularities and mechanisms and show how they have configured differently in different dominant nations at different times. In this way, our materio-spatial understanding can support comparative analysis across material types and across time. Materio-spatial analysis provides a means to compare conjunctures across the sequentially cumulative changes in technology, production, and institutional density—both of state-firm relations in specific social formations and of the world-system itself.*

Concretely and practically, attention to space and matter refines the empirical and theoretical bases not just of environmental history and sociology but of a broad range of other social, political, and economic fields of research. Technical economies of scale and increases in capital sunk in transport infrastructure reduce the cost of moving raw materials through space and so expand the space across which cheap and bulky raw materials can economically be moved. The Japanese restructuring and globalizing of raw materials markets reiterated and expanded earlier technological revolutions—led first by the Portuguese, then by the Dutch, followed by the British and then the Americans—that caused huge jumps in the economies of scale involved in transporting and processing the bulkiest and cheapest raw materials. In the next chapter, we explore the patterns and similarities in the ways that each of these nations adapted their technologies and their political, economic, and financial institutions to the exigencies of matter and space, at the distances and volumes that the scale of production required, to dominate the world economy as it existed at the time each nation initiated its technological revolutions.

In the rest of this book, we will take the lessons that the sequence of extractive economies in the Amazon have taught us about matter, space, and technology and use them to analyze the origins and sequence of the industrial economies that dominated trade with the Amazon. We will see whether we can extend the dynamic interactions of matter, space, technology, and society that are so eminently salient in the sequence of extractive economies in the Amazon to typify global processes and consequences of material intensification and spatial expansion.

Reinterpreting our analysis of local ecological and economic systems in the Amazon in light of global processes and consequences of material intensification and spatial expansion will help us to trace historical changes in the ways that technology mediates between society and nature. We believe that under-

---

*Comparison based on site-specific intersections of materio-spatial regularities allows trans-temporal comparisons that supersede Millian strictures that require control of all but one variable. Such comparison allows configurational (Tilly 1995b) and conjunctural (Paige 1999) analysis in ways that overcome Paige's objections to the social-scientific search for regularity and generality of explanation.

standing these changes will help us to understand the sequentially cumulative cycles that have globalized the world economy.

We will consider how the tension between economies of scale and the cost of space creates technical and organizational challenges whose solutions foment social, political, economic, and technological innovations aimed at overcoming the cost of space as it threatens to eliminate the economies of scale. To do this, we will pay close attention to the ways that scale-increasing and strength-enhancing technological innovations have supported the ascent of different material economies to world-trade dominance.

The technologies that enabled these individual ascents have also driven the sequentially cumulative economies of scale in production and transport over the past six hundred years. Implementation and extension of these innovations expanded the productive and commercial space and the transformation of matter in each systemic cycle of accumulation (cf. Arrighi 1994). The long-term sequential cumulation of these transformations constitutes the underlying mechanism of globalization.

CHAPTER THREE

Between Nature and Society

How Technology Drives Globalization

Contrasting the direct dependence of extractive economies on natural production in single bounded ecosystems with the diffuse dependence of productive economies on multiple extractive economies in many different ecosystems illustrates how the reciprocal determinacies of nature and society shape the material intensification and spatial expansion that drive globalization. The extraordinary global diversity of material configurations of space and of spatial configurations of matter presents challenges and opportunities around which particular nations devised innovations—technological, financial, political, and commercial—that enabled them to dominate world trade for a time. These innovations in turn shaped demand and prices for specific local products available in specific places. Local societies in some of those places reorganized themselves to satisfy that demand, in the process transforming their local ecosystems.

In this chapter, we search for the origins and accumulation of economic and political power in the nations that rose to trade dominance. Each of these nations devised technologies and organizations that enabled them to exploit the bio- and geo-diversity of natural production more effectively, more cheaply, and over a larger portion of the globe than their competitors could. These nations created new social institutions—technological, financial, political, and commercial—to secure privileged access to and exploitation of the diverse raw materials they used. Their control of abundant low-cost raw materials then enabled them to strengthen and expand these institutions and to use them for international dominance. They then used this dominance to expand their access to and lower the cost of even more raw materials.

This required articulation of ecology and economy, and of nature and society, across the three levels of reality we discussed earlier. These nations devised their technological and organizational innovations within the contingent human agency of the third level of reality, but their ability to do so depended absolutely on the diversity of material and energetic forms and the regularity of material processes at the second level of reality. Abstract and constant laws of physics and the rules of chemistry and biology at the first level of reality made these material processes regular and thus knowable and predictable. They also made them diverse and thus humanly useful. Finally, they simultaneously enabled and constrained these processes.

The historically accumulating power and prosperity that technological and organizational innovations made possible depended absolutely on the diversity of material and energetic forms that had evolved through multiple configurations of these laws and rules, but the incorporation of these diverse forms into social history was driven by human agency. In that sense, the ways that technology mediates between society and nature, as well as the ways it mediates the relation between extractive and productive economies, incorporate all three levels of reality. Our goal in this chapter is to explain how and why the intersection of these three levels of reality led to the historical sequence of nations that dominated global trade and to the accumulating increase of technologically, financially, and politically unequal power in that sequence.

We do this by reading matter and space back into the history of dominant economies. We incorporate theories of globalization, written largely at the third level of reality, into a more precise analysis of the ecological and physical realities that shape human history. This enables us to analyze how the intersection of material and social processes made possible the technological change and resulting economies of scale by which a succession of trade-dominant nations progressively globalized the world economy.

World-systems theory has produced some robust analyses of the long-term material intensification and spatial expansion of capitalist economies. Arrighi

(1994), for example, shows that each successive hegemonic cycle or systemic cycle of accumulation over the past eight hundred years significantly intensified material production across expanded commercial space. Each cycle has built upon, and thus expanded, the material and spatial scale of previous cycles. Arrighi, however, attributes the dynamics of these cycles to finance and politics. He claims that capitalists invest in newly developing economies in order to overcome the falling rates of profit that result from accumulating excess investment in the mature hegemonic economy. He acknowledges, but does not explain, the ways that each cycle of expanded production thus heightened national dependence on raw materials that occur in limited and exhaustible quantities in specific places. He thus narrates, but does not explain, why successive ascendant economies transported vastly increased volumes of more types of raw materials across ever-larger distances or how they maintain and coordinate stable exchange and cargo relations with the sources of these raw materials. He sees spatial expansion solely as a financial solution to overaccumulation of capital in the mature, or trade-dominant, hegemon. He thus does not realize that it occurs first as a material solution to the physical requirements of the expanded production that ascendant nations must achieve before they become trade-dominant.

In this regard, he follows David Harvey's (1983) explanation of "the spatial fix," that is, the geographic expansion of capital investment as a response to the "internal dialectic of capital." For Harvey, capitalists compete with each other by investing increasingly large capitals in increasingly productive but site-specific infrastructure, machines, and plants that over time saturate product markets and are themselves made obsolete by even more productive and capital-intensive new technologies. Capital tends to install these costly new technologies in new frontiers with markets adequate to absorb their expanded output.

Harvey explains the expansion of capitalist commerce exclusively as the spatial fix for the site-specific devaluation of capital. He emphasizes finance and so does not consider that expanded material requirements of increasing economies of scale can only be supplied by incorporating previously remote spaces into the global economy as new extractive peripheries. Arrighi pushes Harvey's partial explanation even further by claiming that finance and politics function autonomously from material production. He thus ignores the interactions of material, spatial, and technological components as particular national economies rise to trade dominance. Like most theorists of hegemony, he explains transitions from one systemic cycle to the next in terms of maturity leading to senescence, devaluation, and decay. He explicitly invokes Braudel's image of autumn as the time when capital has reached its fullest growth, starting its decline at the same time it harvests its mature fruits.

We focus instead on the material, spatial, and technological processes that drive trade dominance and globalization. Transitions from one systemic cycle to the next occur when an ascendant economy adapts technologies developed by the established trade-dominant economy to the larger scales and broader spaces that the ascendant economy's particularly favorable materio-spatial situation makes possible. By enabling us to frame a comprehensive account of historical development within each cycle, this perspective enables us to integrate that account with our explanation of the movement from each cycle to the next. We can show how each trade-dominant economy sowed the seeds of its own decline, not by over-accumulating capital but by developing new technologies so powerful that they eventually enabled rising competitors to exploit even larger volumes of matter in even greater space at even lower unit costs. We thus add spring and summer to the autumn of Braudel and Arrighi, including a full account of each cycle within our analysis of how these cycles intensify and expand in a sequence that leads toward their ultimately global scope.

Trade dominance depends on economies of scale. These require expanded production. Expanded production requires expanded volumes of raw material from larger sources across broader spaces. Importing expanded volumes over greater distances requires larger and more efficient vessels and infrastructure. We trace the interactions between the abstractly universal physical laws that govern the extraordinarily diverse but highly regular materio-spatial processes that determine scale-enhancing technologies and the political, financial, and economic dynamics of different nations' invention and use of these technologies to rise to dominance. We can thus explain spring and summer and not just autumn by examining how each nation achieved the material prerequisites for ascent.

These naturally structured material prerequisites create challenges and opportunities for social innovations in finance and politics. Expanding transport infrastructure presents capital barriers larger than the scale of capital accumulated in the previous scale of production can manage, so building transport systems adequate to the new scale of production requires more complex financial institutions that can combine, coordinate, and deploy in single projects a larger number of separate capitals.

Coordinating more and larger separate capitals generally requires participation by more investors from more localities spread over broader territories. The distances separating these localities progressively attenuate the bonds of kinship, acquaintance, and face-to-face interaction. The willingness of individual capitalists to sacrifice autonomous control over their own capital to anonymous and distant others requires establishment of collectively trusted agencies with adequate power to control dishonest or abusive deviation of capital into purposes other than those agreed upon by all investors.

Historically, this regulation and oversight was first managed by municipal, and then by national, states. To establish the necessary trust, fiscal competence, and legitimate authority, the state had to create legal and bureaucratic systems that expanded the moral universe further and further beyond social bases of kinship and primary interaction. These states became more competent and more powerful as the scale of the systems they managed and of the capitals these systems required grew larger to accommodate ever-greater economies of scale.

As we trace the material intensification and spatial expansion of the globalizing world economy from fourteenth-century Mediterranean city-states through the Dutch union of city-states under the Stadt General to the national British state that administered an empire and on to the American invention of an unprecedentedly large republic federating multiple local states, we will also see how financial institutions grew from (1) a base in local noble families in the Mediterranean to (2) a municipally authorized and regulated Amsterdam bank whose letters of exchange were accepted internationally in the seventeenth century to (3) the national bank of Britain that standardized, stabilized, and guaranteed an internationally accepted currency in the eighteenth century to (4) the combination of American national banks, local banks, boards of trade, and stock exchanges capable of uniting literally thousands of separate capitals and deploying them to underwrite the vast and vertically and horizontally integrated corporations that dominated the twentieth-century economy to (5) international, multilateral financial institutions such as the World Bank, various continental development banks, and the IMF, which could underwrite and coordinate the banks of multiple core nations in globally dimensioned financing in the late twentieth and early twenty-first centuries. These progressively larger and territorially more inclusive financial institutions enable each rising nation in turn to expand and strengthen its access and its transport systems such that it can provide more voluminous flows of matter from more, and more distant, extractive economies to its own expanding heavy industry.

This process catalyzed the expansion and strengthening of both finance institutions and the capacity of states to regulate them. Each expansion of transport and finance, in turn, increased the power and wealth of trade-dominant nations relative to their raw-material suppliers.

Capital's financial power increased as it drew on the accumulating profits from the expanded production that this additional infrastructure made possible, but the infrastructure had to be financed before production could expand. Arrighi is correct that finance and politics combine at the height of national trade dominance, but he is wrong to think that either is autonomous from material process. The world economy expands toward globalization not only as a spatial fix for capital over-accumulated in a mature dominant economy searching for the higher rates of profit in less mature economies; even more decisively, it also expands earlier as a material fix for the expanded consumption of matter and energy driven by economies of scale.

In *World Trade since 1431*, Peter Hugill (1994) partially corrects this failure to theorize how material expansion drives spatial extension. He takes as his point of departure an affirmation that Harvey elaborates from Marx—that technology mediates between human society and nature—to analyze the history of technical innovations that have intensified material process and expanded the spatial extent of the world economy. Hugill thus explicitly embraces the historical material aspects of political economy. Unfortunately, he does not systematically explore the financial and political innovations and collaborations that have enabled particular nations to initiate and direct this intensification and expansion to their own advantage.

In the sections that follow, we combine Arrighi's focus on finance and politics with Hugill's focus on technology and space within our three-level analysis of how the accumulating sequence of trade-dominant nations' technological, financial, and political innovations drove and shaped globalization. This enables us to explain the links between technological innovation, trade dominance, globalization, increasing international inequalities, the accelerating disruption of local ecosystems, and the impoverishment and destabilization of extractive economies.

Matter, Space, Technology, and Finance: A Synthetic Model of National Ascent to World Trade Dominance

Dominating world trade requires prior and ongoing social, financial, and technological adaptations to the opportunities and obstacles presented by the intersection of the material and spatial features of the ascendant economy's location with the productive and commercial organization of the world economy at the historical moment of its ascent. In this section, we elaborate a set of linked propositions that we will examine in comparisons of each nation's ascent to trade dominance.

1. National struggles and competitions—military and economic—to dominate trade have driven globalization for at least six centuries.

2. Firms and states introduce technical economies of scale that reduce unit costs of raw material and of labor but that require (a) greater volumes of more types of raw material and (b) more capital invested in larger vehicles, machines and infrastructure.

3. Accommodating the new technology's extra weight, size, strength, and raw material consumption creates a series of challenges whose solution depends on increased political and financial capacity. Expanded transport systems depend on unitary control of extensive tracts of contiguous land for ports, rail lines, roads, warehouses, cranes, and chutes. These require large investments with slow rates of return that benefit multiple competing firms irrespective of whether they share in the investment. The free rider problem must be solved, the agglomeration of property rights must be carried out, and large capitals must be inflexibly sunk into a built environment before the expansion of production can even begin. Solving these challenges expands the functions and the powers both of the national state and of national financial institutions.

4. As economies of scale increase the size of capital required to construct vessels and to build environments on a scale sufficiently large to be cost-competitive, the new forms and scales of finance required for these investments generate innovations that make financial institutions more powerful, with greater subsidization by, closer collaboration with, and more influence on, the state. This effect increases as the growing distance between extraction and production heightens both the need and the rewards of the state-firm-finance collaboration. In addition to collaborating on the construction of technology and infrastructure, states, firms, and financial institutions in the ascendant economy collaborate to devolve a greater proportion of the cost of constructing and administering extractive and transport infrastruc-

ture on the firms and states of the raw materials periphery. Negotiating skills and knowledge in international and domestic relations complement and reinforce each other.

5. Success in these various projects cheapens and stabilizes access to raw materials. Cheap and stable supplies of raw materials provide additional competitive advantages in constructing transport vessels and infrastructure. Reduced transport costs set in motion feedback loops that provide a generalized trading advantage, because (a) the competitive edge in producing vessels for export and for the national carrying trade, combined with the cheapened access to raw materials that efficient transport provides, accelerates throughput, particularly in the shipping and later rail industries (Unger 1978; Chandler 1977); (b) accelerated throughput enhances opportunities for standardization and other innovations that further cheapen the cost of transport; and (c) as the cost of transport falls and the spatial extent of transport infrastructure expands, trade of both raw materials and finished goods into and out of the ascendant national economy's territory becomes cheaper and thus more competitive. Historically, the technological innovations driving the feedback loops that enhance the ascendant economy's commercial and political power have increased the size, weight, and speed of ships, trains, and trucks, reduced the fuel they consume per ton-mile, and extended their infrastructure into new parts of the globe.

6. The reduction of transport costs extends the distances across which it is economically viable to transport raw materials, both those most voluminously used with the lowest value-to-volume ratios and those more valuable specialty materials like rubber, copper, tin, manganese, aluminum, cobalt, and titanium. Reduced transport costs enable the incorporation of greater amounts of increasing varieties of materials from more, and more distant, places on the globe. Expanded and intensified use of transport, production, and warehousing infrastructure further reduces the unit cost of transport and handling and increases the returns to the capital sunk in this infrastructure, enhancing the ascendant economy's trade advantage.

To the extent that over-accumulation of capital stimulates investment in production, rather than in extraction and transport, outside the national territory (cf. Harvey 1983; Arrighi 1994), the greatest opportunities for profit will occur along the transport lines established to cheapen the import of raw materials. Thus, the same mechanisms set in motion to solve the contradictions between scale and space simultaneously globalize raw materials markets; broaden, deepen, and diversify technological innovations; increase industrial consumption of raw materials; and globalize trade in finished goods.

7. Each resolution of the contradiction between scale and space requires further material intensification and spatial expansion. This repeated, cumulatively expanded reiteration of crisis-solution-crisis exacerbates the tension between competitive social drives, such as those toward trade dominance, which work through economies of scale, agglomeration, and spatial concentration, and the limits on natural production.

8. The increasing transport efficiencies underlying globalization steadily reduce the possibility that major suppliers of raw materials will rise to trade dominance, or even to successful industrialization. Each of the technical and social organizational increases in economies of scale sharpen the inequalities between raw materials suppliers and industrial consumers and accelerate the depletion of the suppliers' natural resources. The combined savings in capital and time they achieve contribute to globalization while enhancing the inequalities between the dominant economies and the nations that supply their raw materials.

Raw Materials, Generative Sectors, and National Development

The challenges and the opportunities presented by the basic raw materials industries and by the transport systems on which they depend foster what we call "generative sectors." These are sectors that, beyond creating the linkages that underlie the concept of a leading sector, also stimulate a broad range of technical skills and learning along with formal institutions designed and funded to promote them, vast and diversified instrumental knowledge held by interdependent specialists about the rest of the world, financial institutions adapted to the requirements of large sunk costs in a variety of social and political contexts, and specific formal and informal relations (a) between firms, sectors, and the state, (b) in the form of legal distinctions between public and private, and (c) between different levels of public jurisdiction.

These generative sectors have been as specific to each hegemonic cycle as was the raw materials challenge that fostered them, but they can be conceptualized in terms of the technology, learning, institutional change, organization of firms and sectors, and state formation that allow comparison across cases and over time. They also allow examination of how changes in the world-system progressively expanded and complicated the requirements for national ascent relative to other economies. In other words, they correspond to McMichael's (1990) multiple form of incorporated comparison.

Generative sectors will be more numerous, more easily observed, and more efficacious in those national economies that are growing so rapidly that they must achieve massive increases in the throughput and transformation of raw materials. The concept is relational, however, within a world-systems perspective, thus implying that generative sectors in a rising economy will have significant consequences for economies that export raw materials or trade in other kinds of goods.

The "generative sector" concept can also be used for synchronically framed comparisons or McMichael's singular form of incorporated comparison, that is, comparison between national economies in the same hegemonic cycle. Presence, absence, material base, density of connections, and extent of connections can be specified and then compared for different economies. The idea of such comparisons between competitors for core status seems relatively straightforward. A more relational question would involve looking at raw materials–exporting countries in terms of generative sectors. Terry Karl's (1997) analysis of the impact of oil on the formation of Venezuela's state and economy or Michael Shafer's (1994) examination of how sectors form the state, for example, could both benefit from our explanation based on the kinds of mechanisms specified in the idea of generative sectors. We can also apply the concept to cases of distorted, failed, or impeded development—from Sweden's abortive sixteenth-century attempt to dominate European trade to Germany's similar failure in the late nineteenth century and Japan's in the early twentieth. Or we can use it to compare the different outcomes of heavy industrial programs in, say, Korea—where they succeeded brilliantly—and India—where they failed. It will also help us explain how raw materials exporters were affected by Japan's raw materials strategies.

We begin our analysis of national economic development and economic ascent by focusing on the emergence of hegemonic potential rather than on its maturity and decline. We posit that the beginnings of economic ascent require successful coordination of internal or domestic technological advances, particularly in heavy industry and transport, with the external establishment of access to cheap and steady sources of the raw materials used for heavy industry. The raw materials used in the greatest volume present the greatest challenge and the best opportunity for economies of scale. These economies of scale, however, drive a contradictory increase in transport cost, as the closest reserves of raw materials are depleted as the scale of the industrial transformation increases. The tension of this contradiction between the economies of scale and the cost of space foments technological innovation in transport—in the form of vessels, loaders, ports, rails, and so on; in the chemical and mechanical means of reducing component inputs per unit of output (for example, for coal and iron in steel); and in the control of heat, pressure, and the mix of chemicals that make the unit material inputs stronger and thus enable smaller, lighter amounts to perform the same work. All of these technological fixes drive each other, and all of them tend to generate increases of scale, thus exacerbating over the long term the very contradiction between scale and space that they are designed to solve. The national economies that have most successfully initiated technological and organizational solutions of this contradiction have simultaneously generated their own rise to economic dominance, restructured the mechanisms and dynamics of systemic and hierarchic accumulation, and expanded and intensified the commercial arena of raw materials trade and transport.

We build on leading sector theory (Modelski and Thompson 1996) and on Chandler's (1977) notion of template to posit that the generative sector emerges around the most complex challenge facing rapidly rising economies. A generative sector, however, is not necessarily the sector in which profits are highest. Rather, as Rostow (1960) argued, it is the sector that stimulates innovation and development across other sectors. Cooperation, and therefore trust, between firms, sectors, and the state are essential to overcoming the costly contradiction between economies of scale and diseconomies of space. Moreover, cooperation and trust have become more essential at each successive stage in the evolution of the world economy because the complexity and scale of the social and natural material processes have increased over time.

Generative sectors are conceptualized much more broadly than simply as leading economic sectors. Generative sectors are simultaneously key centers of capital accumulation, bases for a series of linked industries, sources of technological and organizational innovations that spread to other sectors, models for firms and for state-sector-firm relations in other sectors, and catalysts for innovative and more broadly encompassing political and financial relations.

Because generative sectors are rooted in raw materials and transport industries, they have driven economic ascent throughout the history of the capitalist world-economy in core economies and have simultaneously underdeveloped peripheral regions both within and external to the national boundaries of these rising core economies. The linkages from these generative sectors spread throughout the ascendant economy, supplying direct inputs for other industries at lower cost, providing infrastructure available for use by other industries and consumers, serving as profit centers, generating capital for investment in existing and new industries, providing markets for other industries, stimulating the development of capital markets that are then available to other industry sectors, and shaping the general pattern of relations between firms and the state. The myriad linkages, direct and indirect, from the generative sector lower raw materials costs, increase labor productivity, and improve international competitiveness in many sectors of the ascendant economy via this range of linkages.

The tension between the contradictions of scale and space is sequentially cumulative, so each systemic cycle of accumulation has confronted more complex tasks, requiring greater and more efficacious state participation, promotion, and protection, together with more and greater coordination of firms and sharing of both the costs and the benefits of technological innovation within and across sectors (even if they remain competitive for market share). These internal dynamics must also reduce the costs of the raw materials and of the transport infrastructure in the external zones from which they are exported. The cumulatively sequential increases in the scale of the raw material transformation and in the size and capacity of the transport vessels and infrastructure correspond to and make economically viable the expansion of the practical commercial space in each systemic cycle of accumulation. The technological developments generated in response to the contradiction between scale and space for the most voluminously used raw materials provide part of the impulses that create, expand, and restructure the world-system as a series of punctuated cumulative sequences. Figures 3.1–3.3 at the end of this chapter show the huge increases in volumes and sizes of iron extracted, steel processed, ships loaded, and distances carried over the past 130 years.

This solution crystallizes the ways that intersections between the social and the natural, and between the three realms of reality, drive globalization. The increased size and capacity depend both on new social discovery and uses of naturally produced matter and on adaptations of the technologies that incorporate that matter to larger, broader spaces. Thus, commerce in the most voluminously traded raw materials—from wood and grain to iron ore, coal, and petroleum—has proceeded from river-based to lake- and railroad-based to ocean-based transport through the Dutch, British, and American systemic cycles of accumulation and Japan's successful restructuring of these trades into truly global sourcing. Each step of this expansion allows and employs huge increases of scale in transport technologies; larger, stronger, faster ships and trains, and the rails and ports that support them, must be built of lighter, stronger raw materials. Once built, these enlarged, more efficient transport systems allow access to a broader range of raw material sources over broader space.

The introduction of new scales of transport and of industrial transformation, by broadening the sources of raw materials from river basin to continental and then to global networks and thus bringing more raw materials sources into competition for the same markets, systematically reduces the ground rents (see Ricardo 1817; Marx 1894; Coronil 1997) available to the resource-rich economies that export them while increasing their costs. Thus, interactions between scale, scope, and technological innovation, along with denser political and material relations between firms, sectors, and the state increase the productivity, the profitability, and the financial and political power in the national economies that initiate, regulate, and structure each systemic cycle of accumulation. At the same time, these interactions lower the rents to and increase the infrastructural investments of raw materials–exporting economies. Each hegemonic cycle increased the commercially integrated space and the movement of raw materials in this space in ways that exacerbated global inequalities between raw materials exporters and raw materials importers (see Ciccantell and Bunker 1999).

Our concept of generative sector thus extends and refines Rostow's (1960) fundamentally economic notion of leading sector. Generative sectors drive technological, financial, organizational, and political relations, stimulating cooperation across firms, sectors, and states in strategies and actions both domestic and international. The technological advances fomented within the generative sectors follow both forward and backward linkages (cf. Hirschman 1958), most importantly by providing templates (cf. Chandler 1977) for direct application to other sectors, which directly or indirectly constitute clusters or linked nodes in the chains of production (Marx 1894; Schumpeter 1934). Historically, those sectors with the densest forward and backward linkages to other sectors involve the most voluminously used raw materials.

Historically as an empirical process and chemically or logically as a material process, technical advances in fuel efficiency and in the strength of these materials and in their ability to be transported consistently created cumulative sequences toward greater scale. Heat and pressure both become more economical in larger containers, and higher heats and pressures create chemical transformation and mechanical energy more efficiently (Landes 1969).

A historical materialism focused on the mechanisms underlying generative sectors facilitates comparative methods appropriate to the cumulatively sequential processes of a spatially expanding and intensifying world-system. McMichael's (1990, 1992) "incorporating comparisons," Tilly's (1995a, b) "encompassing comparisons," and Tomich's (1994) "commodity circuits" all assume larger and potentially global systemic unities within which comparable instances, differences, or simply cases may occur. All three authors confront the problem of how to compare phenomena that may be linked to each other through various complex causalities of relations, but none (a) problematizes the explanatory status of the larger systemic unities, (b) acknowledges or offers a means to account for the different degrees of intensity or for the significance of the relationship between the instance and the larger processes of which they are an instance, or (c) addresses the explanatory status of the mechanisms that constitute the complex causalities that link instances to the larger systemic unities. We propose that matter and space, as naturally given aspects of physical reality, manifest themselves socially and economically in built or manipulated environments as cost, scale, and distance.

In these and related manifestations, matter and space pose regular, specifiable conditions of production and exchange. Once specified, the conditions may reveal their explanatory status and the intensity of their links both to the local and temporal particularities of instances and to the temporally evolving global systems in which they participate and which they partially form. In other words, we propose that comparison based in highly specified physically and spatially grounded material analysis resolves some of the problems in recent comparisons of instances that participate in complex systems of highly dense interaction, especially when the system itself evolves over time, driven by and driving changes in its component parts.

Since the second industrial revolution—since the growth of economies that build machines to make machines—steel is the most voluminously used raw material, and its major inputs have included coal, iron, and oil. One of the major sites for the social incorporation of these materials has been in the means of integrating space and matter, that is, in the means and infrastructure of transport, which themselves serve most significantly to cheapen the spaces across which these voluminously used materials are transported. Reducing the cost of space requires new and complex understandings of a broader range of materials. These include systematic knowledge of their pure instances, of their transformation into energy, and of their reaction to and incorporation into each other under different temperature and pressure conditions. The social processes of production depend fundamentally on matter, and production-enhancing technologies entwine comprehensively with the historically accumulating social knowledge of and the capacity to manipulate ever-more precise differentiations between the chemical and physical properties of different material forms.

Space defines and organizes the world economy as a system because of the ways that matter is distributed in and across space. Different kinds of matter are located in different places. As technology advances, material forms used for particular production processes or for particular products become progressively more specific. The locations of specific kinds of materials correspondingly become more rare, so that the total distance between the locus of production and the locus of extraction increases. Thus, space and matter are integrally entwined in both production and extraction. Expanded production consumes more matter across broader spaces, so the expanding interaction of scale and distance of matter and space drive the expansion and the intensification of the world-system.

Space is simultaneously a means of production, a condition of production, a barrier to production, a cost of production, and an obstacle to circulation of commodities. Space impinges on extraction even more directly than on production, as the space in which the resource extracted occurs is geologically and hydrologically determined. The attributes of this space include not simply location on a two-dimensional plane, but also the topographical characteristics of the site—of the entire space between the site of extraction and the site of transformation—and the amount of space across and within which a given amount of the resource occurs. For example, in minerals, space is reduced to a percentage of pure ore and a measure of overburden, that is, to the amount of other matter in whatever space must be excavated to extract a given amount of the mineral in question.

The composition—hardness, friability, moisture, and so on—of the surrounding matter combines with this space to determine the costs of extraction and processing as well as the environmental impacts of the extraction. Thus, the relevant space of matter—or, to put it another way, the space that matters—in extraction includes the depth and extent of one form of matter within other forms of matter (i.e., the ground) as well as the naturally determined distances between the sites of natural occurrence and of social transformation. It is in the reduction of the cost of this space that expanded production generates large and complex technological innovations in material and energetic forms that permit increased economies of scale in transport vehicles, loaders, and infrastructure.

Marx (1894), Mandel (1975), Innis (1956), Landes (1969), Chandler (1965, 1977), and Harvey (1983) have all in different ways explained the multiple and complex links between expanded production, technological advances in material use and in energy capture and containment, and new means of transport. Marx (1894), Innis (1956), and Harvey (1983) have all noted the high cost of the building of the environment required for rail and shipping and the role of the state and high finance in overcoming the inadequacies of individual capitals or private ownership of land. Though the role of raw materials procurement and transport and the physically determined or technical economies of scale in heavy industry are consistently undertheorized by all of these authors, the instances in which they have discovered and then presented these relationships of capital and innovation consistently involve the movement of matter across space and the questions of property in both matter and space.

This confluence of space and matter in the formation, expansion, and intensification of the world-system demands a specific focus on the strategies to procure and transport raw materials as these have structured cumulatively sequential systemic cycles of accumulation. The resolution of the contradictions between the scale of transformation and the cost of space has created generative sectors in all of the economies that have become serious candidates for hegemonic status. The material processes and physical attributes of the raw materials and their extraction and transport can be specified in precise, regular and commensurable—and thus comparable—terms theoretically independent of any of the social processes that constitute a relational analysis of the world economy or a comparison of its component parts. We can explain their links to the generative sectors that drive the expansion and reorganization of the world-system. Their explanatory status can thus be quite high.

The semantic form of the notion "generative sector" connotes that other sectors, or changes in those sectors, are generated. Chandler (1977) demonstrated that the railroads adapted to the problems and opportunities of moving multiple cargos over great distances at high speeds by devising organizational forms that could coordinate, dispatch, and monitor entire trains, and eventually individual cars, from and to multiple locations. In his analysis, the telegraph companies, faced with the need to coordinate and regulate the flows of multiple messages from and to multiple locations, adopted the "template" that the railroads had developed. We note, though, that Chandler did not consider how the telegraph, thus formed, increased the efficiency, safety, complexity, and profits of railroads. Our model of generative sectors, therefore, includes consideration of feedback, or autocatalytic loops, between generative sectors and the sectors that they foment or stimulate to change. If raw materials access strategies require new scales of transport—and thus new technologies, new design, and new infrastructure—we expect that successful access to cheaper iron ore and coal, for example, will both require and generate technical solutions to transport costs. The subsequent design and construction of larger ships, better railways, larger ports, more capacious warehouses, and so on will in turn create demand for more steel, and potentially for higher quality steel, in volumes, or at levels of throughput, that both enable technological innovation and allow full use of the capacity required to reduce unit costs of steel. Reduced unit costs of steel make shipbuilding cheaper and more competitive and further reduce unit costs of iron and coal transport. Feedback loops of this type cheapened, deepened, and accelerated the synergies between iron ore, coal, railroads, ships, and smelters after the adoption of Bessemer smelting in the Great Lakes steel belt every bit as much as they cheapened, deepened, and accelerated the synergies between iron ore, coal, computerized continuous casting and hot rolling, vastly larger harbors, the construction and deployment of huge dry-bulk carriers, and the relocation of heavy industries and their dependent consumer industries around deep-water ports in Japan a century later.

In this sense, iron and coal may be treated as generative sectors, but transport is not simply a generated sector. Rather it participates in feedback or autocatalytic loops that intensify the growth and innovation in the generative sector, that is, in the steel sector or in the iron and coal sector. Whether or not the state becomes involved in financing, promoting, relocating, and regulating these sectors, and whether or not it develops skills, creates capable agencies, and inspires trust, are empirically accessible questions about the generative sector's contribution to historically particular processes of national development. The progression toward increasingly tight coupling at greater technical economies of scale across greater spaces and multiple sovereignties with greater levels of production—a progression that appears to characterize the unfolding of the world-system—also suggests, though, that only economies that develop effective generative sectors can rise to economic dominance or competitiveness. These generative sectors only emerge effectively in nations whose physical and social conditions give them a relative advantage in confronting the challenges and exploiting the opportunities that the world-system presents at particular historical moments.

Within this approach, the world-system must be the object of analysis, and global processes are data, but we strongly affirm, against the wisdom of Immanuel Wallerstein and many of the proponents of globalization, that the nation and the nation-state must remain a key unit of analysis. The complex interactions and dynamic processes that underlie the creation and the impact of generative sectors occur within bounded spaces; however much these generative sectors draw on, act within, and affect global systems, we cannot understand how they develop and change unless we locate our analysis in the places they occur. We hope that our insistence that process and location are intimately related may enrich world-systemic analysis of globalization by providing insights into how the internal processes of national development are intertwined with the ways that this development affects the global economy.

Matter, Space, and the Future of Global Economies

The strategies of successive ascendant economies to cheapen transport technology and infrastructure, together with their strategies to stabilize their access to international raw materials sources, have been a major factor in progress toward globalization for at least six hundred years. Systematic analysis of regularities manifest in material needs and of the spatial obstacles to their satisfaction reveals abstractly similar patterns of successful national strategies over this period. We constructed a model of the ways that transport and production technologies have mediated between society and nature—both material and spatial—as social manipulations of nature have expanded and intensified. We use this model in the following chapters to guide our comparative history of trade-dominant nations' reorganization of world markets to achieve cheap and stable access to raw materials. This history of material intensification and spatial expansion will raise a critical question about our future: how and whether a global system that has sustained its competition for markets and materials through successive waves of expansion can continue as it reaches the material and spatial limits of the globe.

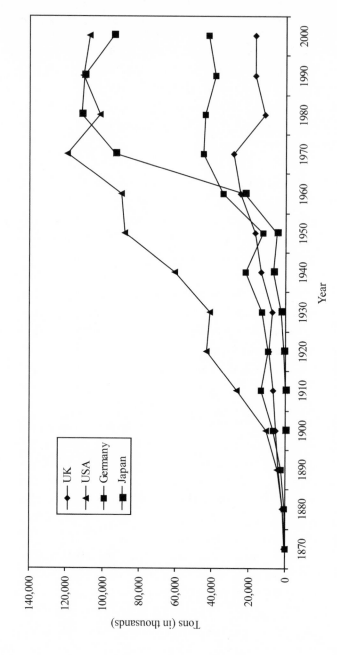

Figure 3.1. Crude Steel Production for the United Kingdom, the United States, Germany, and Japan, 1870–2000.
Sources: IISI; Mitchell (1998)

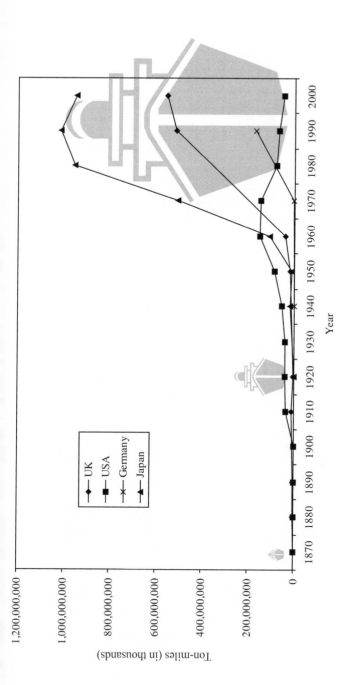

Figure 3.2. Iron Ore-per-Ton Mileage and Ship Size for the United Kingdom, the United States, Germany, and Japan, 1870–2000.

Note: The icons along the horizontal axis indicate the size of a prominent dry-bulk carrier for three points in time. The height of the icon roughly indicates the relative capacity of each ship.

In 1870, the *R. J. Hackett* hauled 1,100 tons of iron ore through the Great Lakes; in 1918 the *Candiope* moved 74,000 tons of iron ore around western Europe; in 1986 the *Berge Stahl* shipped 350,000 tons of iron ore from Brazil to Japan.

Sources: AISA (1902); AISI (various years); BISF (1951); Couper (1983); *Encyclopedia Americana* (1995); Ewart and Fullard (1972); *Hammond World Atlas* (1971); Hogan (1971); IISI; JISF; Maddison (2002); Manners (1971); Marshall (1995); Mitchell (1998); NISF (1923); NSC (1973); Swank (1892); Thompson (1994); USGS/U.S. Bureau of Mines (various years); Warren (1975).

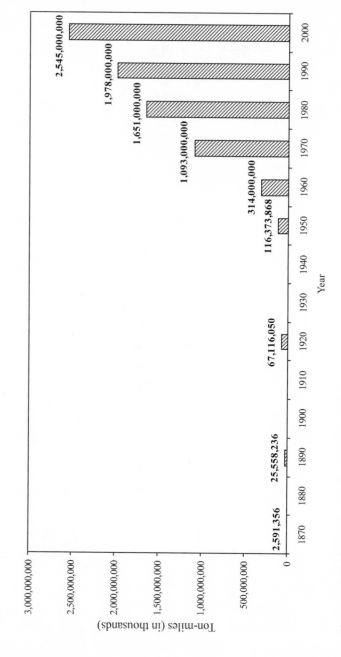

Figure 3.3. World Movement of Iron Ore, 1870–2000.
Sources: Couper (1983); Ewart and Fullard (1972); Fearnleys (various years); *Hammond World Atlas* (1971); IISI, Mitchell (1998).

CHAPTER FOUR

Bulky Goods and Industrial Organization in Early Capitalism

In the following chapters, we apply the model we have developed in our materio-spatial analysis of the Amazon and other extractive peripheries to a comparative analysis of the ways that each ascendant nation expanded the then-prevailing modes of extraction, transport, production, and exchange to achieve trade dominance. Our story starts as Portugal responded to the rich fifteenth-century markets of Genoa and Venice with major innovations in navigation and shipping. It proceeds with abstractly similar but much larger processes in sixteenth- to twentieth-century Holland, England, and America, ending with a consideration of the similarities and differences in Japan's twentieth-century continuation of innovations in navigation and shipping. We will be asking whether the sequentially cumulative effects of this intensification and expansion have resulted in what we now call globalization.

As we saw in the last chapter, it is possible to detect abstract patterns of similarity in the material and spatial processes by which different countries expanded and cheapened their access to raw materials. These abstract patterns and the ways that they have shaped the actual history of globalization emerge from a consideration of particular stories of particular nations at specific times in history. To understand fully the ways that their particular social organization responded to their particular materio-spatial contexts, we must consider how general physical laws and the multiple material processes they structure provided the geo- and bio-diversity that enabled these particular national histories. We will show how each ascendant economy adapted its technologies and social institutions to both its local and its global materio-spatial situations sufficiently well to dominate world trade for a period. We will also show how their successes moved the entire world economy toward globalization and toward greater inequality.

In the present chapter, we focus on competition for trade dominance at the beginnings of capitalist enterprise, when for the first time large quantities of wood, then the most voluminously used raw material, were transported and traded across what were then—socially, economically, and scientifically—very great distances. We will examine Dutch and British strategies to assure a cheap, plentiful, and stable supply of high-grade timber to build the ships that first won and later sustained their trade dominance from the sixteenth through the eighteenth centuries.

For Portugal, protection fees from Genoa and Venice eventually generated sustained financial and institutional support from the crown that built on Portugal's seafaring traditions and felicitous location to foment the technologies and navigational skills needed to couple the sheltered natural environment and prosperous consumer markets of the Mediterranean with the far broader trade opportunities of the Atlantic. Inventing the three-masted ship, adopting and then perfecting the Chinese compass, and figuring out how to mount iron cannon on ship decks enabled the Portuguese to open and then dominate the new maritime trade routes to Africa and to East and South Asia.

The routes through the Atlantic replaced the overland trade routes that Genoa and Venice, in the eastern Mediterranean, had dominated until the Ottoman Empire rose to challenge them. Portugal's innovations in maritime transport vastly expanded the material volumes of this trade; even small sailboats carried cargo more cheaply than horses and camels. Portuguese ships soon turned westward across the Atlantic, incorporating new space and new materials—American sugar, tobacco, turtle oil, and spices, as well as silver and gold—into the Mediterranean-based world trade in Asian spices, silks, and drugs.

Genoa's and Venice's demands for naval protection and for carriage in the Mediterranean of bulkier, cheaper trade goods provided the initial motivation, and later patronage and protection, for Portugal's seafaring commercial development. Portugal's geographic location—where the Mediterranean meets the Atlantic—enhanced the commercial and political opportunities of this patronage. Adventurous entrepreneurs and a supportive, trade-oriented crown familiar with Chinese maritime technological accomplishments provided the local energy and capital for innovations in navigation and exploration. Eventually, the interaction between the natural and social features of Portugal's location within the contemporary world-system allowed it to dominate the most recent transport technologies, control the richest trade routes, and establish access to highly prized commodities—tobacco and sugar—that rewarded crossing the greater transatlantic distances to their sources. Portugal advanced marine technology enormously, but ships were still rudimentary enough to limit Portuguese long-distance trade to luxury goods.

Portuguese and Spanish maritime expansion, and the gold, silver, tobacco, and sugar it introduced into European commerce, generated demand for more and higher-quality materials for shipbuilding and port construction, as well as for grain and fish, than the Mediterranean could supply. Amsterdam's location on low-lying flatlands down the Rhine, the Elbe, and the Weser from some of the densest and oldest oak forests and richest agricultural lands in Europe allowed cheap access to and transport of high-quality timber and wheat. Privileged access to high-quality wood, abundant grain and cheap bulk transport sustained early development of Holland as a provider of high-volume, low-value raw materials to Spain and Portugal.

Meeting the Challenges of Space and Bulk

Because transport technologies are so critical, so costly, and so complex, developing them competitively requires the coordinated collective efforts of firms, sectors, financial agencies, and the state. The potential rewards of trade dominance are huge, however, so firms, sectors, the state, and financial agencies—in a nation where sufficiently favorable materio-spatial conditions are combined with a viable or adequately adaptable base of political-economic organization—may be induced to sacrifice some of their autonomy to enable and institutionalize effective collaboration. Success in such an undertaking strengthens both the state and the national economy.

Procurement and transport of raw materials therefore tend to become generative sectors during the period in which the rising economy restructures the world economy to its own advantage and, in the process, restructures itself. The synergies between the timber trade, shipbuilding, finance, shipping, more diversified commerce, industrial development, and state economic policy in sixteenth- and seventeenth-century Amsterdam, the city that led the Dutch ascent to trade dominance, were rooted in the physical and spatial characteristics of Holland's specific location within Europe's continental and oceanic environment. These physical characteristics formed the defining context for the social and economic organization that the Dutch developed and used to create a highly competitive transport system. This transport system supplied raw materials, merchandise, and capital to interdependent systems of production, finance, exchange, and warehousing. Supported by increasingly efficient and profitable transport, these systems led a national economy that soon dominated world trade.

The technological and international trade context within which Dutch geography and social organization proved so advantageous was particular to the historical time in which Dutch cities began their economic ascent. Capitalist industry, based on large-scale and expansive production of commodities in order to sell them for profit in competitive markets, was just starting its revolution of the world economy. The material needs of capitalist industry for inputs and to feed its workforce cheaply were creating, for the first time in history, sustained and growing demand for cheap transport of bulky matter across large spaces.

Transporting luxury items, including the gold, silver, indigo, and sugar they were appropriating in Africa and America, was far more profitable, so Spain and Portugal relegated the bulk transport of cheap staple goods—for which their overseas enterprises had both created the demand and provided the gold and silver to buy—to the Dutch. Recent changes in climate and demographic changes resulting from the Black Death coincided with agricultural innovations such as three-crop rotations and the moldboard plow to increase the volume and scale of grain agriculture and to extend it farther north than had previously been possible. The conversion of northern spaces from forest to field expanded and cheapened potential supplies of both grain and timber. The Dutch were ideally located, both in relation to the natural northern environments that provided the means of production for wheat and wood and to the social southern environment of the rich and voracious Mediterranean markets, to dominate the newly emergent bulk trade for cheap goods.

In the context of sixteenth- and seventeenth-century technologies and trade routes, Europe's geography privileged Holland's access both to the raw materials, particularly timber, required for shipbuilding and to the most prosperous and profitable markets then available. Geography thus provided the natural conditions for transport to make the rest of its economy the wealthiest and most powerful in the world. As we will see, the other trade-dominant nations also started from naturally and socially created advantages specific to their respective moments in world economic history. Because the world economy changes over time in scale, scope, material intensity, and technological capability, the conditions that make particular materio-spatial and social organizational attributes favorable also change over time.

The location of raw materials and the topographical and hydrological avenues of and obstacles to transport occur naturally in geographical space, but the location and vibrancy of markets and the political power that controls access to them and profits from them are created by social struggles over the matter, space, and human labor that provide these markets with commodities and supplies. During the sixteenth century, Holland and Britain created an alliance to free themselves from their economic and political subordination to Spain and Portugal. They then struggled with each other through the seventeenth and eighteenth centuries to dominate old and to establish new trade routes. These expanding networks first connected resource-rich northern Europe and its surrounding seas back to the Mediterranean, whose recent history as the world's most profitable market had left a legacy of overdeveloped commerce, overextended empires, and overexploited environments. They then incorporated more distant spaces—African, Asian, and American—that the Spanish and Portuguese transport technologies, colonial conquest, and rich-trade strategies had recently included in the world economy's commerce. In order to assess how and how much local configurations of matter and space affected the form and outcome of these struggles and then to assess whether and how these struggles produced the material intensification and spatial expansion that drove globalization, we reanalyze the detailed and extensive existing studies of the origins of trade dominance in Holland and England in terms of the interactions between social organization and materio-spatial features.

Most of these studies examine Dutch and British history to answer theoretically derived questions about finance and politics in national economic development with respect to international relations and warfare, imperialism, and the establishment, maintenance, and decline of hegemony under capitalism. The Dutch and British cases are particularly germane to these important questions. Their struggles to achieve and sustain trade dominance coincided with the origins and early evolution of capitalism. Dutch and British contributions to, dominance over, and successful exploitation of this new and developing system of production and exchange make them the most relevant cases for theorizing both the origins and the abstract economic dynamics of capitalism.

The dynamics of matter and space that were so evident in the Amazon, and therefore in the existing literature on the Amazon's economic history, are somewhat more obscure in the general history of capitalism. We can overcome this drawback, however, by exploiting the same advantage we had in the Amazon, of being able to focus on the ideographic detail of the particular single cases—Dutch, British, American, and Japanese—and reinterpreting that data through the materio-spatial logic we have developed. Some case studies of particular national economic histories are extremely useful in this regard. Some modern Dutch historians, for example, are so fascinated by early Dutch modifications of the environment that they present extensive data on the ways that the environment affected industry and commerce. Though the materio-spatial aspects of these cases have not been much theorized, enough of the relevant information is included in the different case studies that we can pull it together and try to theorize it ourselves.

How Trees and Transport Contributed to Amsterdam's Development

A generative sector is not necessarily the sector in which profits are highest. It is rather the sector that stimulates innovation and development across other sectors. Failure to keep this distinction clear leads to confusing results. Wallerstein (1982) is not alone in thinking that because rates of return were highest in textiles and because Dutch productivity in textiles retarded British development in this sector, the textile trade explains Dutch dominance in the Baltic Sea trade. Barbour (1930), however, shows that Dutch ships dominated traffic through the Baltic Sound long before the textile trade became important. The Baltic hinterlands supplied the high-volume, low-value raw materials—timber, tar, flax, grain—for the vibrant Mediterranean markets that had emerged from the accumulated wealth of Genoa and Venice and the imperial prosperity of Portugal and Spain. The Dutch traded higher-value, lower-volume goods—wine and textiles—from the Mediterranean back to the Baltic region, but their strength in these rich trades lay in their ability to ship less valuable, larger, bulkier materials cheaply and efficiently. The rewards for this ability stimulated further efforts by individual firms, by town and city governments, and finally, by the emerging national state to enhance their already superior transport capabilities.

Both Barbour and Wilson (1973) place shipping and shipbuilding at the center of the complex mix of entrepôt trade, manufacture, finance, and state support that lifted Amsterdam to economic preeminence. Wilson specifically argues that shipping and shipbuilding constituted the major sources of linkages and multipliers as well as the critical source of raw materials: "There seems an incontestable case for arguing that the richest society so far in history had been the creation of sea transport" (329).

From a materio-spatial perspective, though, we can see that Wallerstein and others exaggerate the role of the Baltic trade in the formation of Dutch dominance. Favored access to material shipped downriver from the German and Polish interior provided the initial physical advantages that ultimately allowed the Dutch to dominate the trade to the Baltic region. Analysts of ship construction (e.g., Unger 1978) attribute the Dutch dominance of shipping to their early control over timber coming down the Rhine and then to the proximity to the wood coming down the Elbe and the Weser from German forests. (The Elbe and the Weser flow into the Baltic Sea but to the southwest end, where the Baltic meets the North Sea, a short coastal voyage north of Amsterdam.) Indeed, the Dutch were trading in wood before they started building ships (Parry 1967; Unger 1978). The same trade routes provided them access to flax for sails, to pitch and tar for hulls, and eventually to copper for fittings. This suggests that the critical comparative advantages in the beginnings of Dutch economic development were in the geographically provided control over the river routes to Poland and Germany and to the bountiful fields and forests there.

The Dutch used this advantage to develop a highly sophisticated shipbuilding industry, with cranes, winches, wind-driven sawmills, and other labor-saving devices. The governments of cities and towns financed ports, docks, locks, dams, and spillways that achieved economies of scale by serving many small boatyards without individual ones having to make the large investments such infrastructure would require. Municipal governments also promoted and regulated the guilds that trained specialized carpenters and purchased other labor-sparing tools for shared use by their members in various boatyards. Good tools and experienced craftsmen greatly increased the Dutch competitive advantage. These governments also supported the construction, enlargement, and drainage of the canals that facilitated and cheapened the import and domestic movement of the bulky raw materials of which ships were made. In all of these ways, the state—through subsidies, direct investments, and regulation—contributed importantly to the material, technical, and organizational bases of Dutch supremacy in raw materials procurement and transport.

These advantages accrued slowly from the early fifteenth century, when the Dutch took advantage of the privileged access to German forests and to North Sea fisheries in developing the herring *buss*, which combined seaworthiness, speed, large cargo space, and low manning requirements. The logs, bound together to form rafts for the trip downriver from inland Germany, could carry huge mountains of wheat at little or no additional cost. As Europe's economy started to industrialize, it became increasingly urban. Growing urban workforces needed abundant cheap protein, so a market for long-distance transport of food started to emerge.

Initial Dutch prominence in trade, both overland and by sea, was in wheat and herring. The highly efficient herring *buss* allowed crews to salt and barrel their catch while still at sea. This enabled much longer and more productive voyages. Dutch merchants, anxious to profit from markets that required relatively little capital and imposed a relatively short turnover time, were encouraged and reassured by the city-state's regulation of the sale and management of equity to multiple shareholders. They bought small stakes—sometimes as little as one sixty-fourth of a share—in single fishing boats. Boatyards used their cheap supplies of oak and their wind-driven sawmills to produce these boats cheaply and abundantly.

As the Dutch expanded their herring and grain trades and combined them with a growing trade in logs and sawn lumber, they moved first to re-export Mediterranean wines, then to dye and finish imported textiles, and finally to trade a much more diversified set of industrial products, including bread, refined sugar, crystal, glass lenses, paper, and copper fittings. These different parts of the Dutch economy stimulated, as Wallerstein (1982) put it, "a spiral effect: a circular reinforcement of advantage" (97).

The Baltic trade did not become critical until demand for wheat and timber had outrun the production of the German and Polish forests. By that time, the Dutch had parlayed their river access to cheap and abundant wood into dominance of the shipping trade, including the construction, technology, organization, finance, and capacity required for competitive maritime transport and trade. Their technical, financial, and organizational skills were nurtured by their extensive and profitable commercial activities on the remotest frontiers of capitalism.

The Dutch opening of bulk trade with the Baltic societies constituted the first instance of long-distance trade in industrial inputs and in cheap foodstuffs; it was a major and revolutionary step toward capitalist industry as opposed to mercantile trade. The Dutch thus both gained the advantages and confronted the difficulties of being first movers in a new kind of business. Long-distance trade up through the Portuguese-dominated systemic cycle of accumulation had been limited to luxury items with very high value-to-volume ratios. Even in small amounts, luxury goods compensate trade and transport. Many of them, such as gold ornaments for Inca and Aztec rulers, silks for Chinese aristocrats, and cotton textiles for Indian maharajahs, were already being produced and sold before the advent of European traders. The Europeans' main chore in this trade was to gather small units of high value produced under indigenous technology and organization and ship them home.

Because the volumes shipped were small, the profits realized had to be derived from the high value of the goods themselves rather than from shipping them. These small volumes also meant that there was no particular incentive to increase cargo space by reducing the size of the crew. On the contrary, large crews were necessary to protect against piracy, the danger of which increased with the value and portability of the cargo. The high unit prices of the small volumes limited their potential market, so there was little incentive to build larger fleets of bigger ships.

Bulk transport of cheap goods, in contrast, depends for profit on the efficiency of the transport itself. This requires that particular goods be gathered and loaded quickly and in greater volumes than the indigenous society would have needed or collected at a single port. In order to profit from their transport, Dutch merchants therefore had to organize extraction, production, and transport in diverse places within their peripheral supply zones on scales and with forms of labor relations new to those societies. They then had to coordinate the inland and coastal delivery of materials from their diverse points of origin to a single maritime port for bulk shipping back to Holland or directly to the Mediterranean.

These goods then had to be transported as cheaply as possible, so there was a major incentive to increase available cargo space. Because provisions for crew must be carried on board a ship, the obvious way to increase cargo space is to reduce the size of the crew by designing and building boats that can be loaded and sailed with less labor. The low unit prices of the large volumes expanded their potential market, so there were strong incentives to build larger fleets of bigger ships. Dutch shipbuilders and craftsmen further increased their market for ships by devising new technologies and adapting old ones to increase the productivity of the crew's labor. This reduced the unit costs of the volumes transported even more, which expanded their potential markets even further. These expanded markets increased the numbers of workers employed, the quantities of capital invested, and the profits realized and reinvested in ship-building and in other industries. In this sense, Dutch technological and orga-nizational innovations to enable cheap bulk transport constituted the pioneer instance of the factory system that was to revolutionize European industry, ag-glomerating laborers around machines designed to produce more goods with fewer workers in ways that produced additional capital that could then be in-vested in other, similarly labor-saving and production-cheapening industries.

Dutch innovations in finance, including small equity shares that spread risk among many investors, state regulation and legitimation of banks, and low interest on commercial loans, supported the Dutch capitalists who organized the central and northern European trade in raw materials. Dutch financial institutions were particularly useful in setting up long-term contracts that encouraged landowners with little previous experience of commerce to engage their serfs in planting and harvesting much greater quantities than they had before. The enhanced subordination and exploitation of serfs, many of them recently liberated (Kriedte 1983), was the first instance of a regular pattern in long-distance trade in high-volume, low-value raw materials—that is, trade that exacerbates inequalities within the raw materials–exporting nations as its value-to-volume declines.

In addition to encouraging expanded logging of forests and cropping of wheat, Dutch merchants invested directly in copper and potash mines and participated in their management in multiple remote locations around the Baltic Sea. The skills they acquired in these activities and the institutions in Holland that supported their ability to advance payment for future produc-tion, to pay with reliable letters of credit, and to provide assurance that they would purchase the goods for which they had contracted—all were essential to their ability to reorganize these economies. They used them to persuade their potential local suppliers to organize and manage the production or extraction of grains, logs, and minerals and to bring them from their dispersed sources to the maritime ports in the growing volumes they needed. Their accumulating skills and institutions, combined with high volume, enabled the Dutch to dominate the Baltic region bulk trades even though their geographical location gave them less comparative advantage over that sea than it did over the rivers that connected them to Germany and Poland.

Early access to cheap and abundant supplies of timber underlay the technical and organizational innovations that have so impressed analysts of Dutch shipbuilding. Because Dutch sawmills and shipyards could acquire and process cheap, high-quality timber, they could sell timber products, including ships, very cheaply. The resulting large and growing market for Dutch ships allowed rapid expansion of production. Expanded production allowed the purchase and warehousing of timber and other raw materials in bulk, so that components of multiple different boats could be bought cheaply and then stored for use as soon as a boat was ordered. The costs of producing many ships could be spread across the investments in innovative labor-saving machines and infrastructure, which made possible and profitable investments in wind-driven sawmills, docks, dams, berths, skidways, and large winches and cranes.

High volumes of production also made possible the standardized design and coordination of standardized input supplies, a characteristic of later industrial systems that Dutch boatbuilders invented. Standardization was particularly important for those parts of boat construction that needed pieces of timber that had grown in particular shapes and sizes, such as the stern, the rudderpost, or the hull. Such pieces are rare in nature and only appear as a small portion of large volumes of logs cut, so having them on hand rather than having to search for them in order to fill each individual order allowed substantial savings of time and money.

Standardization and stockpiling thus enhanced the already impressive productivity of Dutch boatyards. Barbour (1930) cites extensive contemporary comments by French and British statesmen about the Dutch speed and economy of construction; these depended not only on labor-saving machinery but also on the ability to manufacture, store, and organize all of the inputs required. In Alfred Chandler's (1965) terms, accelerating throughput contributed the conditions needed to support technical and organizational innovations that in turn created growing economies of scale and speed.

This critical scale of throughput was possible because the Dutch access to timber was so much cheaper than the French or British competitors could achieve. High throughput of wood encouraged the cross-sectoral adaptation of windmill technologies from water pumping to wood sawing (Wallerstein 1982: 97) and the increasing investments of urban governments in shipbuilding infrastructure and equipment and in programs to attract skilled carpenters and shipbuilders. These in turn led to the multiple efficiencies that allowed the Dutch to increase their economies of throughput in building boats for their own expanding and diversifying trade as well as for sale to other nations.

High volume of throughput cheapened shipbuilding and stimulated demand, which in turn increased throughput, fostered technological innovation, and allowed for major state and private expenditure in physical plants and improved tools. These combined results enhanced productivity, lowered costs, and thus increased construction, and therefore volume and speed of throughput, even further. In addition to lowering raw materials costs and allowing for their standardization, the increased volumes of throughput spread the costs of the capital sunk in machinery and infrastructure across more ships produced and sold, thus reducing the capital cost of each ship produced.

The circular self-reinforcement of these linked processes eventually led Dutch shipyards, molded by the evolving interests and accumulated experience of ship owners and shareholders, agents of the city-states, shipbuilders, and ships-carpenters' guilds, to institute some of the earliest forms of industrial organization. The standardization of design and construction allowed for close integration and articulation between multiple suppliers of inputs to ships. It enabled specialized division of tasks and so allowed for the more precise training of labor. The combination of specialized labor permanently employed with the stocking of material inputs allowed the most rapid construction possible (Barbour 1930; Unger 1978). Indeed, volume was so high that processes of standardization could be employed even for the construction of ships adapted to particular cargos and routes (Barbour 1930). All of these latter developments were social-organizational inventions, but all rested fundamentally on the initial natural advantage the Dutch enjoyed in terms of their access to timber.

Matter, Space, and the Dutch Rise to Trade Dominance

Most analysts agree that early Dutch dominance was based on the herring trade (Barbour 1930; Unger 1978; Wallerstein 1982) and that the herring trade depended on the economical construction of ships. They do not, however, fully recognize or theorize how directly economical construction of ships depended on cheap timber, nor how much the cheapness of timber depended on Holland's materio-spatial conditions. They thus neglect the ways that the skills and social institutions—in construction, in finance, and in state-firm collaboration—that the Dutch developed around their materio-spatial advantages in access to raw materials and to trade routes prepared them for eventual dominance of much more complex and diversified carrying trades. In this section, we analyze more specifically how the material properties of different kinds of timber—and the spatial distribution of the trees and the forests that produced them—simultaneously shaped the Dutch economy and its relation to the rest of the world economy.

Albion (1926) claims that the Dutch could build ships for a third of British cost; Barbour (1930) has it that Dutch boats cost half as much. These savings contributed enormously to Dutch dominance in trade, warehousing, and finance as well as to the Dutch position as the preeminent exporters of wood and ships to the rest of the world. They rested on cheap raw materials and high throughput, whence emerged the multiple technical, political, organizational and social innovations that sustained the Dutch carrying industry.

It is well-established that timber in Holland cost half or less as much as in England. Most authors attribute this to the volume and organization of Dutch purchasing in the Baltic region, but it seems likely that the Dutch advantage lay in their privileged access to timber from the Rhine, the Elbe, and the Weser. Loading and unloading long, thick, rigid, bulky logs in vessels with decks sufficiently close to withstand an ocean crossing constituted a large part of the transport costs of timber. Timber rafts coming down the Rhine could be moved along canals and rivers to the mills on the Zaan and in other Dutch towns with no such loading and unloading.

Even where the Dutch did need to load logs into boats, geography reduced their costs. Timber was notoriously costly and dangerous to transport on the open seas because it made boats rigid and therefore susceptible to breaking up against heavy waves. The route from the mouths of the Weser and the Elbe to Amsterdam can be made completely within the sheltered narrow channel between the Frisian Islands and the coast in more capacious, flatter-bottomed, open boats that were easier to load and unload than the deeper-drafted closed boats needed for passage on the open seas. Indeed, Parry (1967) notes that the *fluyt*, the Dutch boat most renowned for its loading, carrying, and manning efficiencies, was an extension of the Dutch inland barges. The Frisian channel, shallow and protected, would have been ideal for this technical extension.

Only Unger (1978) recognizes that the Dutch ascent started while the forests drained by the rivers that emptied south into the west end of the Baltic Sea, that is, the Elbe, the Weser, and the Rhine (which flows through the Netherlands), were still adequate to demand. But even he does not directly acknowledge how much cheaper transport from these rivers would be to Holland than to other European destinations. Albion (1926: 146) mentions rafts of oak eight hundred feet long and sixty feet wide, powered by several hundred rowers, coming down the Rhine to Dort. In the Zaan, whose shipyards and lumber merchants dominated their respective industries from 1600 into the eighteenth century, ships were built primarily of oak from the Black Forest (Van Houtte 1977: 174; see also Parry 1967: 178–79), which must have come down the Rhine.

The special opportunities and challenges that Holland's location and topography offered—in relation to natural sources of high-quality wood and to markets appropriate for its social transformation into ships and other structures—encouraged the Dutch to adapt political and financial institutions they had first developed in order to modify their national environment. The dikes and windmills built to drain their lands and the canals constructed to turn the flat land and abundant water to the service of internal transport had required coordinated efforts, investment, and administration on a territorial scope that exceeded the sovereign control of any particular city-state. The commissions to build and regulate polders, dikes, and canals were thus the first governmental forms to operate on a supra-municipal level. They formed habits, expectations, and a political culture that would later serve for direct municipal, and then supra-municipal, support of the technological innovations and infrastructural investments that transformed Holland's natural geographic advantages into complex political-economic systems. Those systems would help Holland dominate the bulk trade between the Mediterranean, the North Sea, and the Baltic Sea—and in the process make Holland the richest nation on earth, the first republic that incorporated greater space than a single city, and the first modern economy.

The British were so concerned by the competitive advantage this gave the Dutch economy that they added a commercial ban against timber from the Rhine and the Elbe to the general proscriptions that the Navigation Acts imposed on commerce in foreign-owned boats (Albion 1926: 142). Such a specific measure to exclude Dutch commerce from the timber as well as from the timber-carrying trade to Britain confirms the Dutch competitive advantage in this high-bulk trade.

Britain's early industry and navy had benefited from domestic oaks rafted down its rivers, but as boat-building and metallurgy expanded, the demand for wood, together with agricultural needs for cleared land, strained the island's limited supply of oak. The German and Polish forests were far greater and confronted considerably less agricultural pressure.

The Dutch were able to purchase timber more cheaply in Norway and in the rest of the Baltic region than were their British competitors, but the volume and organization of trade that made this possible was economically and organizationally based on the tremendous Dutch advantage of access that timber traders and shipbuilders had to the German and Polish forests. Their sustained dominance depended on the social organization of shipbuilding and raw materials procurement, but this social organization would not have been possible without the tremendous throughput that arose from their privileged access to timber. Even into the eighteenth century, when most shipbuilding timber in the world came from the Baltic region, the Rhine continued to supply the Dutch with cheaply transported high-quality oak (Albion 1926: 147). The Dutch merchants took advantage of the organization and infrastructure developed under the high throughput of cheap German and Polish timber to dominate the trade out of the Baltic Sea as well. Pine masts from Norway became more important as the size of boats increased, so the Baltic hinterland forests did provide critical material inputs, but the mast, by volume, accounts for a relatively small portion of raw materials. Privileged river-based access to oak remained critical to the low-cost, high-volume trade that enabled the Dutch to dominate the Baltic region bulk trades.

Access to Norwegian white pine became more important and more contentious as the British militarized their commercial strategies. Lacking Holland's access to river and coastal routes for timber, the British faced the high cost of loading and unloading ships built in the closed manner needed in the open seas. These additional costs made ships built in England too expensive to compete with Dutch-built ships, so Britain's only path to trade dominance was through naval might. Instead of innovating increasingly cheap ways to build and to operate cargo ships, the British designed and built ships that combined speed and maneuverability with the balance and strength to carry and fire multiple cannons without capsizing. The strong, tall, but light masts that Baltic region white pine provided became even more important as the British attempted to win control over sea trade by designing the world's deadliest fighting ships. Indeed, Albion (1926) reports that gunners targeted masts rather than hulls in sea battles, in part because they were so hard to replace. However important masts were militarily, however, Dutch access to cheaper oak sustained their continued dominance of the carrying trade and of the export of ships long after Britain possessed vastly greater naval power (see Phillips 1990).

A quick look at the map shows that, other than having a coastal voyage along rather than a crossing of the channel, the Dutch do not have a particular locational advantage over England in terms of access to the Baltic Sea itself. The well-documented Dutch advantage in shipping costs within the Baltic region and, indeed, their continued cost advantage in shipbuilding (Barbour 1930) can only have been the result of social organization achieved earlier, from the expanded throughput made possible by their initial but later ended natural locational advantages. Wallerstein's dismissal (1982: 97) of Sir George Downing's worries about the Dutch herring trade requires forgetting what he himself acknowledges on the previous page—that it was the technical innovations in the herring buss that initially made the Dutch so formidably competitive. The *buss*, however, emerged from the cheaper and more easily monopolized river-borne—rather than Baltic Sea—timber. Even after their growing dependence on Baltic region timber eroded the Dutch natural advantage, shipping remained the generative sector. The booming demand for boats in the seventeenth century forced further technical and organizational innovations, which then spilled over into other sectors, both commercial and industrial (Unger 1978).

How Ships and Timber Generated Commercial, Financial, and Political Capacity

Holland's competitive advantage in shipping provided increasing opportunities to build and deploy more ships as well as the basis for public and private support of and participation in a growing variety of carrying trades. Diversification and expansion of these trades encouraged the development of larger and even more cheaply produced and manned boats. By the early seventeenth century, the series of technical innovations in ship design and construction that gave rise to the herring *buss* had culminated in the *fluyt*, the awesomely efficient foundation for the eventual Dutch dominance of the Baltic Sea bulk trades. The efficiencies and productivities in shipbuilding and bulk transport, along with the development of a skilled workforce and labor-saving tools in both industries, provided the capital and knowledge needed for Dutch entry into the traditional rich trades—gold and silver, spices, silks, and sugar—that required longer voyages in larger, more seaworthy ships. Trading companies, organized by the city-states and financed by broad-based subscription in capital shares, extended the commercial arena of Dutch commerce, sending ships in convoys organized by the state and under its military protection (Israel 1989) to purchase high-value, low-volume commodities in Asia, Africa, the Caribbean, and the Americas. There as well, the competitive advantage that cheap construction and navigation of highly efficient ships and the commercial skills and financial agility that they had developed in the Baltic region enabled Dutch merchants to offer better deals to local suppliers than the British, Spanish, or Portuguese colonial powers could match. Lacking the skills, institutions, and access to natural resources that cheapened Dutch transport and enabled Dutch traders to offer better deals, all three competing imperial powers relied instead on armed might to control trade, an expensive and highly unpopular means of striving for trade dominance.

The religious and civil tolerance of the Dutch city-states, in comparison with the religious wars and persecutions of religious and ethnic minorities in the nations with which they were competing, enhanced these material and organizational advantages. Dutch religious tolerance encouraged the immigration of Protestants and Sephardic Jews fleeing from the Habsburg and Iberian inquisitions. Many of these refugees brought capital, commercial experience, and a network of family and community relationships that facilitated long-distance commerce. The Sephardic Jews could operate as effective factors and agents in Spanish and Portuguese colonies, so cultural compatibilities complemented the Dutch carrying trade advantage. As Israel (1989) writes, "The plain fact was that one had a better deal trading with the Dutch. 'The islanders here,' it was reported in Barbados in 1655, 'much desire commerce with strangers [i.e., Dutch and Jews], our English merchants traffiquing to those parts being generally great extortioners'" (244).

Shipbuilding and the timber trade, as generative sectors through the sixteenth century, laid the groundwork for the "moment" of Dutch hegemony between 1625 and 1675 (Wallerstein 1982: 96). The so-called Dutch "mother-trade" to the Baltic region—in herring, grain, timber, tar, flax, and later textiles—that figures so prominently in Wallerstein's analysis of Dutch hegemony was based on the early Dutch innovations in ship design, construction, and manning, which emerged from the early Dutch comparative advantage in access to timber. More importantly, though, shipping was a generative sector in the sense that technical and organizational solutions to the problems of volume and weight created industrial structures well in advance of those in other sectors. Important as textiles were in trade, their form of production was still essentially artisanal or craft-based. Shipbuilding presented problems of weight and volume whose solutions rewarded innovation in technique and in the coordination of public and multiple private capitals, as well as in the division of labor and increasingly in the subordination of guild regulations to economic efficiencies (Unger 1978). Under pressure from and encouraged by shipbuilders and the state, the guilds supported technological innovations, invested in shareable tools and machines, and trained their members in the specialized skills these tools and machines required.

Not only did the various states invest in shareable infrastructure for ship-building, they also recruited and trained labor in the disciplines and techniques required for industrially organized shipbuilding. Dutch shipbuilders were the envy of Europe, and Colbert tried to recruit Dutch laborers to work in French shipyards. Barbour (1930) reports that Colbert was unsuccessful because the Dutch workers feared that, in the absence of the state-supported infrastructure used for moving and lifting bulky inputs and the firm-organized standardization of inputs and divisions of labor, the work would be too heavy. In other words, the famous Dutch labor discipline was based on skill, technology, organization, and state support, not on a willingness to engage in heavy labor. The abundance of cheap, high-quality raw materials enabled Dutch shipyard workers to discover, centuries before it was understood more generally, the potential association between improved conditions of labor and increased labor productivity.

All of these social and economic advances were rooted in Holland's materio-spatial advantages in procuring, transporting, and marketing what were then the world's most voluminously used raw materials. Low construction costs sustained the expanding demand, domestic and international, for Dutch boats. Increasing throughput of raw materials allowed the Dutch economies of scale at the site of raw materials extraction and processing, not only in wood but also in hemp, flax, tar, and pitch.

Cheap transport made possible both the high-volume trade with the Baltic region and the re-export of Baltic hinterland products to the Mediterranean. This high volume enabled Dutch merchants to post resident agents and factors in the Baltic region, compensating for the relatively undeveloped commercial culture there. The continuous commercial presence of Dutch agents on the frontiers of trade expansion facilitated foreign direct investment, as did local merchants' joint ventures with Dutch capital and long-term purchase contracts with Dutch buyers. These exported Dutch institutions facilitated and cheapened Dutch organization of raw materials extraction, local transport, and sale (Lower 1973).

The trade volume that cheap shipping made possible also generated close collaboration between the various city-states and local business. The carrying and marketing of grain and naval supplies, which Barbour (1950) regards as having been "the mainspring of the city's [Amsterdam's] new wealth, as of her earlier modest eminence" required systems of storage and internal transport in Holland as well (26). These in turn required rapid construction of canals, bridges, and warehouses, voracious consumers of raw materials in their own right. These kinds of shared infrastructure could be neither built nor operated without active state participation and regulation. Canals and bridges, essential to the cheap internal movement of goods, were costly to construct and maintain. Warehouses, viable only with efficient transport, were critical to Dutch economic centrality; as Barbour (1950) puts it, "The king of Sweden's most valuable asset, his stock of copper, was piled up in Amsterdam's warehouses" (22). As the general acceptance by foreign merchants and states of Dutch letters of credit depended on Holland's impeccable reputation for effective regulation and fair treatment, so too did the willingness of Dutch and foreign merchants and foreign states to entrust their entire stock in trade to a public warehouse.

All of these functions—credit and finance across broad spaces and long seasons, internal transport, and reliable warehousing—required coordination, collaboration, and recognized divisions of authority and control between private and public capital. The rich rewards available to the state and to private firms from successfully managing these functions made it worthwhile for all participants to accept this division of control and authority. Trade, internal infrastructure, and financial capability all grew together; as they did, Amsterdam became the safest, most convenient, and therefore the preferred market in Europe—both for trade goods and for money. Amsterdam became Europe's entrepôt—goods, services, and money made their way through Amsterdam regardless of where they originated and where they would finally be consumed.

The success of the entrepôt trade rested on the same three functions that Dutch trading strength required—internal transport, reliable warehousing, and an agile but transparent financial system. The Dutch ability to build the necessary infrastructure cheaply and then to finance it expeditiously laid the ground for Dutch investments in foreign mines and smelters for copper, lead, and iron. Access to raw materials and credibility as an entrepôt trader created the conditions for the development of a major munitions industry in the first quarter of the seventeenth century.

Arrighi (1994) explains Dutch profits from the European wars of the seventeenth century as an example of capitalists in a mature hegemon searching for quick profits from finance rather than from the slower material processes of production. In fact, Dutch financing of these wars was directly related to the very material process of manufacturing munitions. The manufacture of munitions required multiple raw materials and multiple skills, and so required the complex organization of extraction, processing, and production in multiple sites. As Barbour (1950) writes, "The Baltic trade brought in potash for making gunpowder, the Italian trade sulphur, and saltpetre could be had from several European countries and from the East Indies" (36). Cannon and ball depended on the organization of smelters and foundries in Sweden and Germany. Sulfur and saltpeter extraction were technically and organizationally less demanding, but they nonetheless required that the Dutch find their sources and stimulate their production and bulking at ports (see also Hobsbawm 1968: 45). Transport efficiencies and the ability to stimulate production in the multiple foreign locations of the inputs required for these products all supported Dutch domestic industries and the highly remunerative trade in provisions and arms to the various warring European nations. Financial agility and flexibility were certainly important in setting up the extraction and transport of materials from their diverse places of origin, but it was materials and material processes, rather than financial speculation, that supported Dutch profits from European wars.

The greatest Dutch profits—from the English civil wars and from the Thirty Years' War—came from the same material processes on which the entire economic Dutch success was built: the timber and grain trades, shipbuilding, and transport. Dutch merchants provisioned grain to English armies in Scotland and Ireland and outfitted Genoese, Venetian, French, and Portuguese navies and merchant fleets during these wars. Here, too, finance and politics facilitated this trade, but Dutch participation in and profits from military provisioning had deeply material roots.

The same materio-spatial logic shows the error of categorizing Dutch textile manufacture as a leading sector simply because it returned high profits. The linkages of other industries to shipbuilding, and the organizational innovations that made it so competitive, were far denser than they were in the textile trade. British commentators of the time and the authors of the Navigation Acts alike (see Albion 1926; Barbour 1930) understood this more clearly than many recent authors. The same British commentators were also keenly aware of the advantages that access to cheap timber and a reputation for fair dealing gave the Dutch.

Much of the Dutch competitive advantage in textiles resulted from technical savings in textile finishing, but in contrast to shipbuilding, textile finishing remained artisanal and had relatively few industrial linkages. Even in this low-volume, high-value sector, transport efficiencies may have contributed significantly. The transport of textiles from Holland to the Baltic region must have been enormously cheap because of the backhaul capacity in the huge Dutch bulk carrying trade from that region. As Barbour (1950) observes, "Grain provided cargoes and paid freights to keep Amsterdam's merchant marine moving, and so made possible cheap transport of commodities less ship-filling in bulk" (27). The Dutch advantage in the Baltic region was for a time extreme; over five months in 1645, "of a total of 1,035 ships which passed through the Sound . . . all but 49 were Dutch" (267n). Dutch exports of raw materials from the Baltic region were the necessary condition for the creation of textile markets (among others) there in the first instance.

Profits from the textile trade were critical to Dutch capital accumulation and to the retardation of British industrial growth. They did not, however, lead the Dutch economy; rather, the textile industry was favored by the industrial and transport efficiencies that rested on cheap raw materials and high levels of throughput in a closely integrated system of supply and haulage. Dutch profits in textiles, as in the other "rich trades" of spices and sugar, depended on the capital, the skills, and the institutions—most directly manifest in the huge fleet and the thousands of specialized workers that built and operated it—that the Dutch had to develop in order to profit from the bulk trades.

Even the eventual profits from finance that so impress Arrighi and other analysts were rooted in Dutch access to material goods, which gave them a material advantage in transport and trade. Barbour (1950) attributes to shipping efficiency and scope the emergence of Amsterdam as the world's principal money market. Money markets can only function in economies that have earned a reputation for predictable and transparent transactions. The densely linked economic and political relations between business and the city-states had been made possible and attractive by the high-volume throughput of raw materials into technology and infrastructure that allowed Holland to dominate world trade. It was precisely the collaborative relations between the public and private sectors that allowed the transparency and predictability required to transform financial institutions designed first to enable Dutch commercial transactions abroad into a money market for all of Europe.

The Dutch merchants benefited in multiple ways from this transformation. Their own ability to extend credit safely to their foreign suppliers was greatly enhanced to the extent that their suppliers used Dutch banks. Participation in the same financial institutions by supplier and purchaser offered recourse to either if the terms of contracts were not fulfilled. These effects enhanced the Dutch banks' domestic and international reputation for reliable and transparent transactions.

They also underlay the Dutch merchants' reputation for fair dealing, generous advances, and high prices, which allowed them to smuggle high-value luxury goods out of colonies controlled by Spain, Portugal, and England in the East and West Indies and the Caribbean (see Israel 1989: 244). These imperial powers lacked the favorable access to timber that had enabled the Dutch to develop the cheapest, most efficient, and most profitable bulk carriage in the world. Nor did they have financial institutions and facilities comparable to those that the Dutch had built from their privileged material base. Instead, they were obliged to resort to expensive military and administrative attempts to restrict the trade of their colonies to their own national merchants and boats. Imposing these restrictions raised the administrative costs of the colony, and the restrictions against cheap Dutch shipping raised the costs of transport. These high costs and the protection the colonial laws gave them against foreign competition lowered the prices that Spanish, Portuguese, and British merchants offered to their colonial suppliers.

The accumulating experience of realizing high profits from high-quality, low-cost construction and management, accessible financial terms, and collaborative relations between state and business had enduring general effects on the institutional and cultural forms that Dutch capitalism developed. Mainstream explanations from Weber (1904–5) to Gorski (1993, 1995) of the original formations of capitalism and of Dutch preeminence in those formations attribute much of the impulse toward capitalism to cultural and cosmological innovations. Recognizing, though, that particular patterns and expectations of social behavior, and the belief systems in which they are rooted, play key roles in any nation's rise to trade dominance does not adequately explain the ways those belief systems emerge in those nations that dominate trade.

Our material-historic approach to the Dutch rise, on the other hand, suggests that culture and cosmology, as socially constructed and transmitted systems of knowledge, evolve in interaction with the environment that the society that creates them exploits to subsist and to profit. The environment does not determine the stories, explanations of general processes, or statements of abstract principles that make up its belief systems, but if the behavior guided by these belief systems is not sufficiently compatible with the environment to achieve its intended results, the belief itself will be changed (cf. Peirce 1877). Similarly, beliefs that support behavior that brings intended results in the surrounding environment will themselves be reinforced and will also generate extensions and refinements around further attempts to exploit or profit from the environment. Dutch technologies of hydrological control, capture of thermal and wind energies, and shipping, as well as the social forms of finance, labor organization, and political regulation, all developed in ongoing interaction with the environment. Each innovation engendered changes in belief systems, and each new belief was subject to the test of utility within the given environment. In this sense, the particular challenges and opportunities that an environment, such as that of Holland, offers will, through ongoing interaction with the human populations that work to modify and exploit it, mold the social construction of belief that provides the basis for social action.

We will explore this idea further in our accounts of how the materio-spatial conditions of each succeeding hegemon shaped both its political and economic institutions and the political culture that underlay them. We will examine especially how increasingly complex and costly manipulations of the national environment in ways intended to facilitate production and trade engendered new belief systems that supported coordination and risk-sharing across ever-broader and more diverse firms, sectors, and state agencies operating in ever-broader territories. The expanded moral universe that enabled these public-private collaborations and acceptance of regulation was clearly cultural, but its origins were found in material struggles to overcome challenges and to reap opportunities within the national and the global environment. Their growth and development were nurtured by their own success, which was also a function of the materio-spatial context that engendered them.

The same public-private collaboration and acceptance of effective state regulation supported the adaptation and extension of another financial service critical to Holland's leadership of expanding world commerce. In 1598, Amsterdam passed a municipal ordinance establishing a chamber to register and settle controversies arising from marine insurance policies. As Barbour (1950) explains, "The chamber earned the confidence of the business community, and successive revisions of the original ordinance attest the increasing volume of insurance registered and the greater precision of the chamber's procedure" (33).

All of these forms of finance—long-term credit, low interest rates, acceptable letters of credit, reliable banks, and trustworthy insurance policies—provide a means of sharing the risks and benefits of expanding commerce into more distant spaces in order to procure the additional raw materials required by the expanded economies of scale. All of these forms of finance depend on impartial state regulation acceptable to and trusted by all parties to any financial transaction. In this sense, effective public-private collaboration to establish financial institutions competent to provide the capital and the security that enable firms to expand their operations materially and spatially is a necessary condition for a nation's ascent to trade dominance. Each nation that has met these conditions has also driven the world economy further toward globalization.

In contrast, though, to Arrighi's claims that finance and politics operate independently of material process as the mature hegemon moves the world economy toward globalization, more widely ranging financial institutions competent to operate on an expanded scale in an expanded space are generated by a national economy's materio-spatial needs at the beginning of its ascent. They then facilitate the material intensification and spatial expansion that drive globalization. In order to serve the expanded economies of scale that drive the world economy toward globalization, these financial institutions must be more highly capitalized, more flexible, and more reliable. They facilitate and expedite globalization, but the fundamental dynamics driving both finance and globalization are eminently material.

As the first national economy to rise to world trade dominance through competitive bulk shipping, Holland was also the first national economy to combine these financial functions effectively at a large scale and high volume. We will see in later chapters how each succeeding hegemon repeated this process on an expanded scale. Each fashioned state-business relations to meet the financial conditions on which its raw materials procurement strategies depended. In each case, expanding raw materials throughput catalyzed and

supported expanded public-private investment in stronger, faster, more extensive transport systems.

These systems resolved the immediate contradiction between economies of
scale and diseconomies of space. Each time, the resulting economic intensification and expansion set the material and spatial context for the next iteration of this contradiction. They also set the technological, political, and commercial context for the next iteration of its solution. Each time, this solution
was initiated by a new rising hegemon introducing a new systemic cycle of
accumulation.

In the next section, we will discuss the peculiarities of Britain's search for
raw materials and compare the consequences of British and Dutch strategies
on each domestic economy and on the world-system. We will analyze how
New England fitted into but also benefited from this process. Then we will turn
to the parallels and differences between these wood-based processes and the
rise of mineral-based heavy industry in nineteenth-century England and
America and postwar Japan.

Timber's Importance to the British Economy

In this section, we describe the paradoxical cycle of raw materials shortage
and procurement that led Britain to a military rather than a commercial
strategy for economic development. Essentially, we argue that, lacking cheap
access to timber, British shipping and shipbuilding could not compete with
Dutch transport and therefore were subordinated to Dutch commerce. Protectionist measures against foreign shipping made timber more expensive, and
military measures against foreign shipping made British ship design and construction more costly and less efficient. At the same time, the military solutions
to the economic problems caused by the timber shortage were themselves
timber-intensive. The British required vast amounts of wood to build their
navy. Naval demand for cannon consumed—as charcoal—a great deal of the
timber available in England. In order to supply grain for the expanding workforce that built ships, smelted iron, and made guns and for the soldiers and
sailors who used them, large areas of forest were converted into farmland. This
conversion diminished replacement of harvested trees. Lacking efficient shipyards and ships, the British either captured ships or imported them from
colonies. Both forms of procurement heightened international tensions and
required more naval construction. In important ways, then, British imperial

conquest and two centuries of continental war resulted from Britain's timber problem—and greatly exacerbated it.

The dynamic mercantile development of Britain was very much the source and the result of the timber problem there. The national shortage of timber drove trade, and trade drove much of production—but much of this production was driven by military rather than commercial demand. Thames-borne trade drove the massive incorporation of raw materials into the built environment of London; much of this material was itself imported. Because timber was so much more expensive to import into Britain than into Holland, British shipbuilding and shipping—already impeded by high raw material costs—could not compete. British economic development was further hampered by Dutch commercial and manufacturing success, because British manufacture and trade were less competitive in European, American, and other markets (Hobsbawm 1968; Davis 1975).

The British state determined to combine a series of protectionist laws with military aggression to maintain the British fleet (Williams 1972). These military strategies aggravated the raw materials shortage that British trade policies had been designed to overcome. Enforcing the Navigation Acts required more armaments. Ships and armaments were also required to protect national shipping and limit the shipping of rivals and to keep open sources of raw materials threatened by reaction against the Acts, such as when Sweden attempted to close the Baltic Sea. Britain's military strategies responded first to the greater efficiency of Dutch shipping and later to continental resistance provoked by Britain's military devices to overcome her own carrying inefficiencies. The rapidly developing iron industry had been almost stagnant until the crown decided to promote the domestic production of cannon through the training of labor and the establishment of smelters (Cipolla 1965). As that industry expanded, so did its consumption of oak, leading to conflict between steelmasters, farmers, and the Royal Navy. The growing costs of administration, including the support of a navy necessary for the security of trade, drove the crown to look for new sources of revenue. The easiest source, in the short term, was the sale of trees from royal forests to the more dynamic industrial interests, especially the ironmasters. In the long term, of course, the sale of oak to the ironmasters simply exacerbated the timber shortage that was contributing to the crown's revenue shortfall. The burning of oak for smelters was a very hot issue through most of the seventeenth century (Albion 1926; Perlin 1991).

Britain's choice of a bellicose strategy to overcome her locational disadvantages in access to timber led to major technical innovations in military vessels and armaments, as well as to state-sponsored piracy in the name of war. It also engendered extreme distortions in the agricultural and industrial economies. The formula that economic dominance is the source of military dominance (cf. Misra and Boswell, n.d.) does not apply to the British case; rather, maritime military might, specifically over the industrially efficient Dutch, was the origin of British economic dominance. Dutch prizes and, later, American-built ships made up a varying but significant amount of British shipping capacity until the adoption of metal-hulled, coal-fueled boats finally gave Britain the raw materials advantage that has underlain all other cases of a single nation dominating world seaborne trade.

The British case is exceptional, but the basis for the exception illuminates general rules. The consequences of the exception are also illuminating: Albion (1926), Barbour (1930), and Davis (1962) all document British complaints, both public and private, about the far greater efficiencies of shipbuilding, first in Holland and then in the American colonies, that made British shipbuilding uncompetitive. Brenner (1993) and O'Hearn (2002) document British intersectoral contention and controversy over colonial and mercantile policy. Where Dutch trade dominance had been rooted in competitive efficiencies, British dominance required trade restrictions that favored some national business sectors at the cost of others. Thus, British external trade policy engendered internal dissent and warfare. In contrast, Dutch external trade policy fitted with internal social institutions devised to modify the national environment in favor of the national economy. While control of the Stadt General engendered controversy and struggle, the state itself developed relatively free of the sectoral efforts to capture state policy.

State and Capital Compete for Wood in Britain

Britain had seriously depleted its forests by the end of the sixteenth century (Albion 1926; Lower 1973; Wallerstein 1980; Chew 1992). In addition to the demands of a growing merchant and naval fleet, several other very widespread uses for wood, including metallurgy and the baking of bricks, construction (such as in the shafts of mines), and wooden barrels and boxes for the transport of trade goods, combined to outrun the regenerative capacity of the island's forests. State attempts to regulate wood use were too little and too late (Evelyne, cited in Merchant 1983). In other words, according to Merchant

(1983), Britain faced serious shortages in the "industry most dependent on wood and most critical to sixteenth century commercial expansion and national supremacy" just as it was embarking on major participation in overseas trade and maritime power (65).

The contradiction between conservation and capital accumulation was already evident by the middle of the sixteenth century. During Elizabeth's reign, a strong nationalist and mercantilist policy developed. Burleigh undertook to develop England's maritime power by encouraging fishing as a "nursery of seamen" (Innis 1956) as well as privateering and by promoting the production of hemp, flax, sailcloth, and timbers (See 1928). He also promoted the repair of ports. Cutting oaks of a foot or more in diameter was prohibited within fourteen miles of navigable rivers in order to assure adequate supplies for naval use, but the state continued to grant cutting rights in order to raise revenues, presumably under the pressure of increasing commercial demand. The royal reserves had been extensively cut, and by 1660 few large forest tracts remained (Albion 1926).

Of the multiple sources of demand for timber, none was so dynamic as the increasing consumption of iron. Much of this demand was directly related to the demand for naval ships. New casting technologies enabled production of much lighter and more accurate and powerful guns and ship's cannons. Britain had imported most of its sophisticated armaments, but Henry VIII pushed for the development of a national armaments industry. The existence of iron ore deposits within the huge oak forests of Sussex determined the choice to locate the industry there.

The amount of wood that was consumed to make a single ton of pig iron was enormous, and oak was preferred because it would burn longer and hotter. The ironmasters came into conflict with local agriculturalists over access to fuel and building materials and with the navy over the destruction of potential shipbuilding materials. Multiple attempts to regulate the ironmasters failed, in part because they were serving critical defense interests. In addition, their use of the forest was a critical source of income to the state, which was perpetually short of the funds required to develop the naval power that depended on the trees being destroyed to make armaments. The chairman of a commission of inquiry that had found that forest destruction by ironmasters had led to riots by farmers was actually beheaded after the leaders of the iron industry protested his findings (Perlin 1991).

Not only was there growing demand for timber, but the quality requirements of the wood were also becoming more rigorous. The simultaneous drive to reduce shipping costs and to increase security motivated great advances in shipping technologies. These technologies made the physical requirements of specific types of timber for masts, rudderposts, keels, and planks more precise.

The Portuguese caravel was an enormous ship, and it had probably reached the limits of size possible with wood construction. It was, however, extremely cumbersome and slow. Military control of the seas favored lighter, faster ships, but they needed sufficient balance to allow long-range cannon (Cipolla 1965). Older military techniques of ramming and boarding with heavy boats gave way to maneuverability with light boats. Both cannons and masts had to be above the water line, thereby tending to overbalance the ship. Ships' timbers had to maximize strength and sail area with as little topside weight as possible. As the importance of cannon increased, British naval design moved toward a combination of speed, maneuverability, hull strength, mast height, and stronger, lighter guns in larger and more complex arrangements. Military demand and military finance stimulated rapid technical development and product improvement in metallurgy, as well as powerful but very expensive technologies of ship construction (Linebaugh 1992). As long as iron smelting used wood, naval demand created conflicting, uneconomical tensions over raw material in the two industries that supplied its most critical needs.

Albion (1926) showed how the engineering and construction of British military vessels required a wide range of different species of wood of precise sizes, especially oak planks and white pine masts. These different types of wood grew best in very different environments, and naval timber needs drove the British to widely distant forests with widely different ecologies and social organizations. The quality of oak was critical for resisting dry rot, and the size, straightness, weight, and height of the white pine was critical for the mast. It is perhaps in the search for masts that British capital and the state most diligently bowed to the weight of nature. Mast height was directly related to speed; mast strength was critical to safety. Reducing mast weight contributed to balance and to ease of repair. British interests would search for appropriate combinations of height, weight, and strength wherever these occurred in nature, incurring huge costs to conquer territory, import labor, regulate property, and negotiate trading rights in order to assure access to the white pine (*pinus strobus*)—the rare, spatially dispersed species that best combined the required characteristics. This meant, essentially, that more and more wood of very precise specifications, and therefore on average more distant, was turned into boats, which were required to transport the huge amounts of wood that were being consumed, much of it in shipbuilding, at the same time that other boats were required to force and secure access to these forests and to repel rival nations that also sought privileged or secure access. Acquiring wood required and consumed growing quantities of wood.

The dynamic interaction of industrial and military demand and the dependence on wood for continued economic growth required that increasing amounts of wood were imported. British entrepreneurs and traders started to establish trading houses in various ports where the large rivers that drained Scandinavia, Germany, and Russia emptied into the Baltic Sea, from which they contracted with individual nobles or with communities for logs. Timber had been imported from Russia since the Middle Ages, but the trade expanded very rapidly at the beginning of the seventeenth century (Lower 1973). In Scandinavia and Germany, this trade eventually developed sawmills as well, where large planks, or deals, were sent to England. The most fundamental changes in social and economic relations in these areas were probably created earlier by the Dutch trade in timber.

Transport was the main cost in wood delivered to England, and it directly affected the forms of processing and utilization of the wood. Logs were squared off in equal width and breadth from top to bottom. Lower (1973) points out that the best wood—that is, the outer part of the tree—was therefore wasted, as was most of the wider base of the tree. If the timber was squared by broadaxe, which it was for most of the early period of the trade, the parts of the tree that were removed were reduced to unusable chips, thus increasing the waste. Transport costs were so important, however, that the amount of occupied deck space and the ease of handling outweighed the timber lost through this manner of preparation. Distance created significant price differentials between different exporting and importing ports. British boatyards were established at points close to the Baltic sources of timber—not necessarily locations conducive to maximal spread effects or other industrial efficiencies (Davis 1962).

The growth of the transport economy and of the navy on which it depended stimulated other economies that required the same raw materials. Access to foreign timber thus became critical for multiple purposes, especially in the dramatic surges of demand such as that occasioned by the London fire—which the British diarist Samuel Pepys characterized as warming Finland's economy. Britain suffered from the limitations of her own supplies and the distance to other sources. Holland had no supplies of her own but was ideally located to exploit the grain and the wood supplies of the Rhine Basin as well as of the Baltic region. The strategies of firms and states, and the relations between them, reflected the very different spatial configurations of raw materials distribution as these interacted with the social organization of raw materials procurement and transport, with very different consequences for each economy.

State, Economy, and Raw Materials in Holland and Britain

Rivers and later canals fed both the Dutch and the British economies, but their different spatial configurations had very different impacts on class and sectoral relations. For the Dutch, economically significant rivers rose in widely dispersed foreign lands rich in raw materials and flowed thence to Holland. For the British, economically significant rivers flowed through the national territory with multiple competing national claimants to raw materials and other natural resources. British economic power in the preindustrial and early industrial period was closely linked to maritime trade and to naval security (both in wooden boats), and to wood- and later coal-fueled metallurgy. In addition to depending on water for the cheap transport of wood, many of the early technical advances in both wood processing and metallurgy depended on water power to drive sawmills and to power the bellows and hammers that increased fuel efficiency and labor productivity in iron-smelting. Agricultural products also moved more cheaply by water, though the savings were not as important as in shipbuilding. The importance of water transport meant, though, that wood and agricultural land near watercourses were highly prized and that the various entrepreneurs who required timber for metallurgy, for shipbuilding, and for a series of other uses competed with each other and with agriculturalists, as well as with representatives of the state, for control over and access to wood. Albion (1926) points out that British naval requirements for timber competed with both corn and iron, and Ashton (1924), in his *Iron and Steel in the Industrial Revolution,* wrote of the "tyranny of wood and water" (22). Part of the problem was that oak, whose tannic acid enhanced its resistance to worms and fungus, grew best on relatively fertile soils. The landowner favored multiple grain crops on a yearly basis, with their greater and earlier revenue, over oak, which would be harvested by his grandchildren. Indeed, the records of intense conflicts between different users of land and raw materials in Britain and the contradictory and transitory attempts by the state to resolve them suggest that the near-total dependence of Japan and Holland on *imported* raw materials relieved the state of irreconcilable demands from different sectors and facilitated the close collaboration between firms and between states and firms in raw materials access strategies.

The naval demand for cannon stimulated the notable British advances in metallurgy (Harris 1988)—certainly the intense competition between naval and metallurgical consumption of oak quickened the search for ways to use coal to smelt iron. The elimination of sulfur from coal-reduced iron ore was a critical step in the later technical achievements that allowed for the cheap production of steel, without which British dominance of shipping and industry could not have occurred. In this sense, it could be argued that the problems of raw materials access set the stage for later British industrial development, but the immediate effects of naval dominance on shipbuilding itself were actually detrimental. The increased speed of construction to which Weber refers was of a very different nature than that which occurred in Holland (Linebaugh 1992).

The Dutch speeded up production by complex division of labor and management of inputs in response to their original advantage in access to critical raw materials. For the British, increasing speed of construction resulted from a financial commitment from the crown, one that could only be sustained by military predations on other shipping powers and by the establishment of empire. Albion (1926) writes that in many cases speed of construction was driven by military emergencies. The conflicts between wood-using sectors and the tremendous drain of the wasteful and costly naval procurement system on crown revenues led to periods where boats were neither soundly built nor adequately prepared. Confronted by a military threat, the crown would allocate resources for construction and repair, but neither shipyards nor wood stores would be adequate on such short notice. Typically, then, insufficiently cured oak would be used, and dry rot would set in early. This process essentially disorganized the shipbuilding industry, raised its costs, and vitiated the quality of its product.

The eventual preeminence of the British shipping and shipbuilding industries over their Dutch counterparts took centuries to attain. It was finally achieved only after (a) preferences enacted by Britain to favor nationally owned boats, (b) a century and a half of war, including massive capture of Dutch boats, and (c) the significant contribution of American-built boats to the British fleet. Even then, the British preeminence was tenuous. Two wars against the North Americans set the New England shipping industry back severely, but by 1860 the Americans were challenging British shipping around the globe. Lacking the Dutch and American advantage of cheap and secure access to timber, Britain depended on a navy, a costly and precarious solution.

Dutch and American efficiencies were difficult to overcome. Under the Navigation Acts, many British shipowners found ways to buy Dutch-built boats and register them as British (Davis 1962). The extremely heavy costs of three maritime wars, commitment to an extended land war, and the refusal of all but the port-possessing provinces of the United Provinces of the Netherlands to contribute to the support of the navy combined to create inflation and a heavy national debt, both of which made Dutch industry and trade less competitive and so contributed to the decline of the Dutch shipping industry (Boxer 1965). Even so, total Dutch carrying capacity and Dutch sale of boats exceeded that of Britain well into the eighteenth century and was only surpassed during a period when Britain was not only buying a significant number of boats from America but could count boats owned by colonial shippers as part of her own fleet. By this time as well, Britain was confronting a major maritime challenge from the French. Britain did not manage to dominate the world carrying trade until after 1814; it achieved this preeminence by success in the European wars and in two wars in America, which not only destroyed rival shipping but restricted access to British and colonial ports. U.S. shipping prospered despite these restrictions, and the embargo against U.S. ships became impossible to maintain as the rapidly expanding British textile industry became more and more dependent on American cotton (Morison 1941). U.S. shipping capacity nearly overtook—some authors claim actually did overtake —that of Britain by 1860. As Mathias (1969) put it, "The heart of the matter here was design, where Britain was 20 years behind. For their size, the Americans were manned with fewer men, they sailed faster, and they carried more. To the British chagrin, even Lloyd's gave them a lower insurance rating" (312).

How Access to Raw Materials Changed Britain's Position in the World-System

We have argued that Britain's shipyards were inefficient because (1) the high price of timber prevented effective British competition in bulk trades, (2) this in turn hindered the development of scale-dependent technologies to cheapen construction and handling, (3) the response of the British state and of British firms to their inability to compete in terms of carrying efficiency was to compete militarily, (4) the military and political rather than economic or cost-reducing forms therefore dominated boat design and shipyard organization, and (5) these military forms produced boat designs that emphasized strength and maneuverability rather than capacity, ease of loading and unloading, and minimal crew size. War with Holland, war in the Baltic region, and struggles to control American forests, together with attempts to inhibit other nations from carrying goods to England, all revolved around military and colonial attempts to overcome raw materials access problems. Britain's access strategies dramatically restructured the economic and social organization of multiple core and peripheral regions, but the economic distortions and waste that resulted from military and protectionist strategies prevented the progressive synergies that developed around raw materials access and transport in Holland. Hobsbawm (1968) insists on the precarious condition of Britain's industrial dominance until the 1840s; his descriptions contrast vividly with the integrated growth and prosperity that Barbour (1950), Van Houtte (1977), and others portray in Holland.

British access to critical raw materials depended on a navy that was first predatory and then imperial. That navy greatly expanded British needs for raw materials, and was thus both an instrument and a cause of military and colonial conquests. Captured and illegally purchased Dutch ships compensated for some of the British carrying disadvantage until the smaller, cheaper, smaller-crewed ships from North America took their place. Davis (1962) estimates that as much as a sixth of the British fleet was American-built, and a far larger number of American-owned ships carried raw materials to Britain until 1776. Deane (1965: 114) counts shipping and shipbuilding, along with textiles and metallurgy, among the three leading sectors of the British economy from the 1750s through the 1830s. Hobsbawm (1968), however, claims between 1800 to 1860 French boats were better built and American boats much better built and that only protective preferences sustained the British shipyards. Far from serving as a template for industrial organization as the Dutch shipbuilding industry did, the British shipbuilding industry distorted the allocation of capital and labor, exacerbated inter-sectoral tensions in the domestic economy, drained state revenues, and increased the costs of imported raw materials.

The stimulus of British military demand for iron and steel on the one hand and for ships on the other had very different effects on the long-term efficiencies and developmental linkages of the two industries. Britain led the world in iron and steel production until the 1880s and remained strong in that sector into the twentieth century. The innovations of British ironmasters are myriad: they discovered the sulfur-reducing chemistry required to smelt iron with coal; progressively reduced coal charges per unit of production; developed the Bessemer converter in 1856, making mass production of steel possible; invented the Siemens-Martin open hearth furnace, which increased productivity; and widened the range of ores from which steel could be made with the Gilchrist-Thomas basic process (Isard 1948; Hobsbawm 1968; Harris 1988). Each of these innovations reduced costs of production, improved the product, or both. They thus extended and cheapened the use of iron and steel across a broad range of industrial processes in ways that tremendously facilitated the incorporation of metals and coal into the technologies that enabled Britain to lead the Industrial Revolution. The chain of inventions that enabled these innovations was set in motion and sustained by the well-financed military encouragement of technological improvements in metallurgy to make possible lighter, stronger armaments.

The effects of demand for ships were dramatically different. Military demand for warships promoted speed, balance, and maneuverability, all essential qualities for warfare but only marginally important for carriage. Britain's boatyards excelled in naval design, but naval contracts did not foster cheap or efficient construction any more than naval design encouraged carrying efficiency. Politically based preferences and specifications encumbered both the already disadvantaged procurement of raw materials and the use of labor (Albion 1926; Linebaugh 1992). British-built wooden ships were thus fast and highly maneuverable but expensive to construct and inefficient for hauling cargo. British shipping was only competitive under a series of costly protective preferences within the context of a cumbersome empire. British shipbuilders— capitalists and workmen alike—increasingly chose to emigrate to America, where cheap and abundant high-quality wood reduced both capital cost and labor effort and enhanced returns to both (Morison 1941; Perlman 1934). Military demand for iron and steel production fostered and strengthened domestic industry, but, as we will see in the next chapter, military demand for timber and for wooden ships stimulated the development of an overseas industry that led first to a war for independence from Britain and then contributed to the rise of American hegemony.

These different outcomes of military demand for different products can be traced directly to the very different materials from which they were made. Wood procurement was a major factor in Britain's relations with European nations and in her colonial strategies. Davis (1962) writes that shipbuilding requirements shaped colonial strategies. Albion (1926) and Lower (1973) both claim that British timber needs directly molded colonial administration and production as well as the establishment of trading houses in the Baltic region and in Russia. Britain's raw materials procurement strategies directly affected both domestic industry and international relations, but in very different ways and with very different consequences than in Holland. Both were intimately linked to the ecological situation of each economy, and both had very different consequences for the other parts of the world whose economies were structured by these two rising economies.

We have shown how the distinctive materio-spatial conditions of Holland and of Britain interacted with changes in the scale and scope of the prevailing technologies and patterns of production and exchange at different times to generate very different patterns of growth in the two national economies and in their relations to the rest of the world-system. Dutch technological and organizational innovations greatly reduced the cost and expanded the practical spatial sphere of bulk transport, thus incorporating previously unincorporated populations into world commerce. The British adapted Dutch transport technology and their own military technology and organization to vastly greater spaces and material needs, decimating native populations and extending their own national populations into the newly opened spaces. Subsequent rising economies continued both of these expansions—improving and cheapening transport across broader spaces to supply raw materials for expanded production.

We will notice as we examine these later cases that, as transport efficiencies increased, transport vessels and infrastructure absorbed a decreasing proportion of the total volume of raw material imported. This happened because their efficiencies increased with economies of scale and speed. An economy of scale in transport occurs because the cargo a vessel can carry increases with the cube of the radius of the hull, while the surface of the hull itself increases only as the square of the same radius. In other words, the material in the hull itself increases much more slowly that the material that hull can carry. The material used in construction and the fuel used in mobilization both increase with the surface of the hull, rather than with the larger increases in the cargo volume.

This ratio can confuse analysis. The point is that, whereas the proportion of total raw materials consumption dedicated to boat and rail construction today is far less than it was in seventeenth-century Amsterdam, the total volume of materials in boat and rail construction is far greater than it was then. Transport constitutes a smaller proportion of industrial production, but it is even more critical to industrial production than it was then, because the much greater variety and volume of raw materials are much more broadly dispersed across the globe. Matter matters, and space has weight, but modern technologies progressively learn to do more with less. Doing more with less increases total profits, total wealth, and total demand, so industrial societies respond to the savings of raw materials by using and consuming even greater volumes of matter (Jevons 2001; Bunker 1996). This expands global production at even greater rates, driving even more extraction from increasingly distant places.

CHAPTER FIVE

From Wood to Steel

British-American Interdependent Expansion across
the Atlantic and around the Globe

Britain's use of her North American colonies to supply her growing seventeenth- and eighteenth-century timber needs set in motion dynamics that vastly extended the space and hugely increased the volumes of raw matter incorporated into the world economy. They also set in motion linked cycles of reciprocal technological innovations. Americans adapted British technological innovations to the far greater economies of scale that America's vast spaces and huge material resources made possible. The Americans' improvements on British technology enabled them to expand and intensify their extraction, transport, and trade of raw materials, both for their own and for Britain's rapidly growing industrial economies. Sustained expansion and diversification of raw materials supplies enabled the British to develop further technological, as well as political and financial, innovations at home, which the Americans again adapted and enlarged to fit their own materials and space.

These cycles shaped the expanding world economy, the hierarchy of nations within it, and the configurations of matter, space, and technology that molded globalization. They developed around shifting disjunctures between production and transport technologies, in which new production technologies required more and cheaper raw materials than existing transport technologies could deliver economically. Such disjunctures consistently emerge in the development of any technical solution to the contradiction between the economies of scale and the diseconomies of space, but the Atlantic was a far greater space than the world economy had ever attempted to straddle. Numerous innovations over many years were needed to enable transport technology to deliver large bulk loads cheaply across the ocean or between the American coast and the continent's vast interior. A vast resource base and scarce expensive labor encouraged the invention and use of labor-saving tools and machines.

The combined opportunities and challenges presented by the vast and bountiful American landscape fostered and then supported increased size, complexity, and competence of technological, financial, managerial, political, and labor organization. The political culture and economy of the original city-states on the Atlantic coast evolved in response to the new materio-spatial conditions and to the technical and organizational adaptations they required as the population and economy moved west. They expanded first as small farms and craft shops over the Appalachians and Alleghenies into the Ohio River Valley and on to the Mississippi and then as larger and larger corporate ventures—first trading for fur but soon speculating in canals, railroads, mines, and land farther north and on across the Great Plains. They then moved on to the huge southwestern and western territories wrested from Mexico and, finally, to extensive trade, and some imperial conquest, in the Pacific Ocean, whose vast spaces were far more remote than the Atlantic from European competition.

The expanded spatial challenges and material opportunities in each phase of this westward drive expanded capital's scale, scope, and power—and that of the state. The political culture's roots in small-scale agriculture and manufacture, though, enabled labor, both hired in industry and producing on its own land, to organize effectively. Labor's insistence on its rights to a fair share of what it produced worked both to expand domestic markets and to bound abuses of power by capital and the state. It thus contributed importantly to the uniquely successful emergence of the United States from the role of extractive adjunct to that of hegemon. We will see that, like much else in American history, labor's social and political power was rooted in the uniquely favorable ways that matter, space, technology, finance, and politics entwined over time.

As American manufacture increased agricultural and extractive productivity thanks to railroads and steamboats, which reduced the costs of transport to the coast, and steel ships, which reduced the costs of transport across the ocean, increased low value-added imports from the United States enabled British industry to move to higher value-added production. In ways that were never duplicated, the flow of matter across great space, from one economy to another, accelerated the industrial development and inter-sectoral articulations in exporting and importing nation alike. The accelerated sequence of new technologies and economies of scale, combined with America's extraordinary ecological diversity and resource abundance, molded and sustained the richest, most complex, and most consequential interdependence in history between an established hegemon and its primary raw materials supplier.

Because American success occurred with such exceptional speed, depth, and duration, it has been broadly theorized into abstract, ahistorical models of social and economic development. In this abstract form, the British-American experience has served as a template for mainstream theories of national development. Core nations invoke numerous variants of these theories to legitimate the extractive economies they promote in peripheral nations and to justify their claim that extraction drives industrial development. Sustained conjoint and simultaneous industrial growth in a dominant productive economy and in its major raw materials supplier is historically unique and economically anomalous, but theorists of national development have posited both experiences as ideal types. Core nation states, firms, banks, and development agencies today invoke the past American experience of raw materials extraction and export in order to secure the acquiescence and complicity of peripheral nations in cheap and stable core access to their natural resources (Bunker 1989b, 1994a). By presenting unique and temporally specific events from the past as unchanging laws, these arguments, and the theories on which they are based, obscure and mystify the material, technological, and political dynamics that increasingly impoverish peripheral extractive economies and ecologies.

In order to demystify these dynamics, we will examine how uniquely the available technologies and the possible innovations over three centuries of development intertwined with the size and the material attributes of the different spaces the American economy confronted and exploited. Extraction drives industrial development only under the most exceptional of circumstances, and the expanding scales of extraction and industry make this outcome less and less likely. The close and sustained intertwining of raw materials extraction, technological innovation, and industrial growth in America resulted from specific configurations of matter, space, technology, and time. Attempts to replicate this experience in other places and at other times are likely to fail and, in the process of failing, are likely to accelerate globalization at the cost of exacerbating its inequalities. In the rest of this chapter, we will examine the extraordinary material and spatial dynamics and processes that sustained this conjoint growth and the intense globalizing pressures that British-American interdependence created in the world economy.

How Matter, Space, and Geopolitics Shaped New England Shipping

The complementary and interdependent economic development of America and Britain commenced while New England was a colony and continued as an independent United States developed the technologies, the transport infrastructure, and the financial institutions to become the most rapidly developing industrial power in the world. Paradoxically, the primary and sustaining function through the first 250 years of American industrial development was to facilitate—and to increase profits from—the export of its raw materials.

Britain's attempts to solve her timber shortage for naval construction extended the trade and transport of its bulk goods across the Atlantic Ocean. The natural sources of these materials were dispersed over a far broader space. The space between Europe and America was far greater, and far less sheltered, than the pioneering Dutch bulk trades had traversed. The Atlantic offered a much more difficult and dangerous challenge to bulk carriage than had the Rhine or the access to the Baltic along the Frisian channel.

The British built strong, fast, maneuverable warships. Transport technology, though, had advanced little since the Dutch developed the *fluyt*, so the greater distance from raw material sources imposed proportionally greater costs on industry. As a result, there were strong incentives to construct ships close to the American sources of timber rather than in the British ports from which they would sail. It was far more profitable to send timber from America to Britain in the form of a ship carrying cargo than of cargo taking up valuable space in a ship.

The spatial distribution of various raw materials had made Holland the most convenient, the cheapest, and the most secure place to build Dutch boats. The shipbuilding industry and the sectors linked to it thrived there. England offered no similar advantage, especially once it became dependent on American timber. Admiralty regulations demanded domestic construction of naval vessels, but English demand for merchant ships generated more shipbuilding in America than in England (Albion 1926; Morison 1941). The multiple harbors of New England were more convenient, cheaper, and—given their distance from European rivals—probably more secure than those in England.

New England enjoyed even more remarkable physical and locational advantages for the combined development of the timber and shipping economies than the Dutch had. The American carrying industry was highly dynamic, but it did not at first drive internationally competitive industrial and financial development in the ways that Dutch shipbuilding did, both because it was so distant from European markets and because it could rely on the sophisticated British banking system. The self-reinforcing economic development that led eventually to American dominance of the world economy only occurred after a long turn away from the ocean and expansion inland. Nonetheless, examination of the New England shipbuilding industry allows for instructive comparisons with the Dutch case. It also shows the enduring consequences of British raw materials dependency.

The rise of British maritime power in the eighteenth century reflected not only Britain's growing industrial power and trade dominance but also the national state's use of military power and trade restriction to overcome British disadvantages in boat construction and raw materials transport. The Dutch had used the concatenating series of efficiencies and economies of scale that flowed from their privileged access to cheap, high-quality wood. Their preference for the principles of free movement and trade on the high seas that Grotius had formalized and that the Treaty of Westphalia had concretized was rooted in their competitive advantage in shipping. Lacking this advantage, the British used force to gain trade dominance, in the process generating intense and prolonged conflict with their European rivals, first with Holland and then with France and a shifting coalition of allies. American merchants, manufacturers, loggers, miners, and farmers, first in the colonies and then in the independent nation, benefited from the effects of these struggles, as they increased demand for American products and simultaneously deflected European military forces that might have been used to control American territory.

The growth of international competition against Britain, and particularly the continental system and French-initiated attempts to constrain British trade in European waters, led to increasing exports from the North American colonies (Albion 1926; Lower 1973). The transition was a difficult one, however, as the entrepreneurs on whom British access depended were loath to invest heavily in North America when there was always the possibility that the more accessible Baltic trade might open up again. The British state, therefore, enacted a series of special privileges and incentives to stimulate investment in North American timber (Lower 1973). As that trade expanded, the colonial state was also impelled to encourage migration and settlement to provide labor for the timber camps. The more that the trade of timber and other raw materials to Britain expanded, the greater were the opportunities for the emergence of North American shipbuilding and carriage.

There are remarkable parallels and critical differences between the Dutch and the New England experiences. Both Holland and the American colonies were able to supply timber for shipbuilding as colonies of the dominant economies of the time, and both remained critical as suppliers of raw materials to the economies of their colonizers even after their respective wars of independence. Their colonial or dependent status provided them protection and thus obviated the need for them to develop major naval facilities while each developed its shipbuilding industry and maritime trade.

America had ample supplies of wood to meet both industrial and transport requirements. After the end of the French and Indian War and the expulsion of France from Canada in 1763, both of which had limited and often interrupted access to northern timber, the greater distance from other sovereignties freed America from the direct need for naval protection that afflicted Holland. Britain undertook the elimination of the pirates that so hampered the early West Indies trade (Shepherd and Walton 1972). Britain's assumption of most of the expense of controlling the seas enabled New England shippers to sustain lower gun per ton and man per ton rates than either British or West Indian boats (Davis 1962).

Both Holland and the United States were situated so as to enjoy cheap access to a wide range of timber types. This diversity of supply distinguished them from the various Baltic nations that had attempted to establish dominant transport regimes. Sweden had access to only a few types of timber domestically and was not well situated to carry on a diversified trade. Her ships could be easily bottled up in the Baltic, and she was too far north to serve as a convenient entrepôt. Economical access to herring, grain, pitch, tar, and timber allowed the Dutch to dominate the bulk trades of the Baltic region,* and

*The Dutch called this bulk trade their "mother trade."

inexpensive access to cod, grain, cotton, indigo and timber allowed the United States to become a major merchant power until steel ships ended the wood-and-wind shipping era.

For both Holland and New England, the most important impulse toward mercantile dominance was in the character of wood. Easily transported down rivers, it is very costly to transport on the open seas. Bulky and heavy, it required large ship tonnage to transport, was difficult to load and unload, and made ships both top-heavy and rigid and thus more likely to break up in heavy seas. Insurance and labor were both costly and difficult, especially as ship owners tended to risk only older boats in the trade.

Like the Dutch, the New England shippers could move timber to shipyards without an ocean voyage. Just as it was ultimately far cheaper to build boats in Holland and sail them to Venice, Spain, or Portugal, so construction in Boston for a voyage to London saved the costs of shipping timber overseas. American shipbuilders were also able to exploit their very cheap wood stores; indeed, they benefited indirectly as well from the timber shortage in Britain, as both capitalists and craftsmen migrated from England to take advantage of the opportunities for profit and employment that the abundant timber supplies of the Atlantic coast offered (Beard 1930). The greater distance from Europe created even greater savings for shipbuilding in New England, though it put North America at a disadvantage in relation to the Baltic raw timber trade.

The great and important differences between New England and Holland lay in history and topography. Holland's hinterland had been relatively densely settled for centuries, and its societies were constituted into political units capable of significant defense and aggression. Thus, bellicose expansion of territory was impossible, and war, particularly land war, was extremely costly. Dutch trade in grains and timber depended on, and led to, massive changes in the property, labor, and commercial relations of a long-settled European interior. New England enjoyed a hinterland whose earlier occupants had been severely dislocated and were progressively diminished in number and political strength. The European migrants who displaced them came to America to take advantage of the property, labor, and commercial relations being established there. The American grain and timber merchants could not depend on already established populations to supply them with raw materials, but over time came greater possibilities for economic integration with the settler communities that the North American timber trade spawned (Morison 1941). Incorporation of additional territory into the United States combined raw materials sources and urban industrial centers within the same sovereign unit.

On the other hand, the Dutch enjoyed closer proximity to denser, economically more dynamic markets for more finished, higher-value products such as textiles and wine, and they were able to exploit the huge demand for armaments and provisions during a century of European war. New York, Boston, Salem, and some of the lesser ports served as important transshipment points between, for example, Honduras and Europe, but the distance from other economies restricted their opportunities for developing an entrepôt trade. Slave economies to the south provided some demand for finished products, including cheap cloth and shoes, and rum processing in New England added significant value to exports that gained a world market. Nonetheless, except for the ships that they constructed so well and so cheaply for the European market, the Americans remained essentially provisioners of raw materials or tramp traders par excellence. Proximity to other industrializing nations allowed for deeper, wider, and much more rapid development in Holland, but it also allowed the powerful military and trade coalitions that arose against her.

New England's coastal and interior trade area was about the same size as Holland's grain and timber hinterland, so that the absolute distances between raw materials sources and markets were not strikingly different. Topography, however, hugely favored New England. Holland's timber either came down the Rhine or flowed down multiple rivers to Baltic Sea ports. All of these rivers drained northern ecosystems. The Baltic Sea essentially runs east and west, offering little variation of latitude. Access to southern systems was only possible after a long westward passage through the channel and down the Atlantic coast and was therefore subject to threat from France, Britain, Portugal, and Spain. In the American colonies, the Atlantic coast runs northeast to southwest, parallel to the Appalachians. This meant that a series of rivers runs west to east across a wide range of latitude and temperatures. Oak, spruce, fir, pine, as well as pine pitch and tar, could all easily be transported down these rivers to sea ports, from which coastal shipping could move them between various boat building and trading centers (Morison 1941). The boatyards in Bath, Maine, for example, imported different kinds of wood, as well as hemp (used for sails) and pitch and tar, from various river systems farther south.

Many of these river systems also supported agriculture and mineral extraction, so a diversified domestic trading system could develop at relatively little cost. Access to northern cod greatly stimulated trade, shipbuilding, and the development of maritime skills. As construction and navigation skills developed and as sufficient capital was amassed to sustain the extended voyages required, New England captains and capitalists developed the risky, but highly profitable, trade in the southern sperm whale, which they had discovered to have more, and better quality, oil than the northern whales the Europeans hunted. By 1833, 70,000 people worked in whaling or in linked industries, and by 1844 $120 million of capital was tied up in the industry (Olson 1947).

The ease of trade extended to the Caribbean and the booming sugar industry there. Sugar, as well as the molasses and rum into which it was processed, was a huge consumer of wood for fuel and for barrels. The slaves on the sugar islands provided a large market for fish, particularly the low-quality fish dried and salted in summer that complemented the market for the higher-quality winter catch that was sold in Europe. Thus, northern fish and timber could be shipped south to very different ecological zones with very different products, while internal industry was facilitated by cheap access to a range of intermediate ecological zones along the North American Atlantic coast.

Another tremendous advantage for New England was the abundance of and cheap access to natural resources that were for the most part available without the encumbrance of recognized property rights. North American timber, for example, could compete in the European market despite the much greater distances it had to cross because there were no rents charged by land-owners for its extraction, because the voyage downriver from stump to port was generally shorter, and because there were far fewer intermediaries between stump and ship (Albion 1926). The Baltic trade to England, on the other hand, involved purchasing agents in the major ports; traders, who contracted with these agents and then filled their orders through agreements with landowners; peasants and serfs, who cut the timber; peasants who brought the timber down the long rivers; and, finally, sawmills in the ports, not to mention tariffs in the Sound paid to the Danes.

The most telling difference in the organization of the North American, and particularly the Canadian, timber trade was that in Europe established populations with other forms of livelihood were recruited into timbering on land already owned and valued for other purposes than logging. Lower (1973) reports that lumber traders made contracts, sometimes as much as three years in advance, for delivery of logs at the various ports on the Baltic Sea and then negotiated the harvest with individual landowners, with contracts specified tree by tree. The landowners in Europe were not driving a frontier in search of speculative and extractive profits, and they had a subordinate laboring class under their direct control. There was already an established wheat economy. Rafts of logs sent downriver sometimes also carried wheat, thus enhancing the compatibility of and not the competition between forestry and agriculture. The huge amounts of branches and slabs left after squaring logs were probably used for fuel rather than left as tinder for the kinds of devastating forest fires that affected parts of North America. There were good reasons, therefore, for the European landowners to husband their forests rather than simply to cut them out as quickly as possible and move on.

In much of North America, in contrast, and particularly in Canada after the American Revolution threatened British access to the forests to the south, settlement was driven by the lumber trade, and cutting was conducted by owners or licensees with no other interest in the land. Both because labor was scarce and expensive and because new timber concessions could be had after the original ones were exhausted, North American timbering tended to be far more wasteful and destructive. The abundance of the natural resource outside of established property relations and the scarcity of labor combined to make rapid but wasteful exploitation more profitable. This problem was exacerbated in areas where there was no agriculture, as labor had to be recruited and food imported.

The most critical difference between different timber regions, however, was whether they developed their own boatyards. The Baltic region's timberlands did not develop maritime industries, despite Peter the Great's clear understanding that shipbuilding was essential to industrialization. Perhaps hampered by the annual freezing of the St. Lawrence River, Quebec never developed a dynamic shipbuilding industry but only made rough boats designed for the timber trade. New England, and particularly Maine, however, did develop a major shipbuilding industry, in large part supplied with wood that came cheaply and conveniently down the short Appalachian rivers to good ports along the Atlantic Ocean.

Beard (1930) suggests various linked explanations, starting with the inland drive of independent farmers looking for their own land up the various valleys flowing down from the Appalachians, the relatively inhospitable soils that drove the colonials back toward the sea, and the potential to combine fish and timber trading with the Caribbean sugar industry and the exchange of fish and whale oil with European manufacturing industries: "By the middle of the eighteenth century, New England was launching seventy five new ships every year, New York and Pennsylvania forty five, and the states to the south forty. Already London Shipbuilders beside the Thames had begun to complain that their trade was declining, their workmen migrating, and their profits disappearing as a result of American competition" (91). Privateering and piracy, possible because of the political conflicts in Europe, also stimulated shipbuilding. American ships were excellent, soon the best and fastest in the world.

Similar complaints and fears from other sectors, in particular metallurgy, brought relief from the British colonial state in the form of restrictions on manufactured and high-value exports from the American colonies—wool in 1699, hats in 1732, and iron in 1750 and 1757. In addition, particularly valuable commodities—rice, molasses, and naval stores in 1705, beaver skins and furs in 1721—could be exported to Britain only. Britain also attempted to control colonial consumption of white pine suitable for masts.

Significantly, though, there were no restrictions on shipbuilding in America. British ship owners criticized the short life of American boats, but they nonetheless bought enough of them to constitute up to a sixth of British tonnage (Shepherd and Walton 1972). More important, American colonial-owned boats, considered British under the Navigation Acts, supplied much of the European-bound capacity in bulk goods, particularly timber. Britain's timber problem left her dependent on colonial shipping; the cheaper American craft essentially replaced the captured Dutch craft in compensating for Britain's timber shortage and the consequent inefficiencies in shipbuilding. Unencumbered by the need for a large navy and blessed with continued access to cheap raw materials, the Americans would go on to challenge Britain as the preeminent transport economy.

The combination of available land and unclaimed timber relatively close to ports in New England allowed for a far closer integration between logging, farming, manufacture, and shipping than was possible in Europe. In parts of the American trade, it appears that individual yeomen sometimes cut and delivered timber to water-powered mills or that state-managed cutting parties were contracted (Perlin 1991). The lengthy voyages down European rivers (in some cases two years) were not necessary—indeed, the short steep flow of many of the northern Appalachian rivers obviated the need for rafts altogether. Ownership of logs within these smaller systems could be determined by branding rather than by human escort. More important, the timber-driven agricultural frontier interacted with the shipping industry, providing labor and supplies, goods for export, and a market for imports.

The still-rudimentary technologies available for long-distance bulk transport in the seventeenth century created the initial conditions for American development of industry to support its basic role of raw materials supplier. The vast distances across the Atlantic and within the American continent favored the location of timber-based industries, especially shipbuilding and shipping, close to the American sources of raw materials. Britain's mercantile policies did not recognize the obstacles distance imposed on transport and on direct administration. Despite clear warnings from Adam Smith and a few British parliamentarians, the British state's heavy-handed attempts to restrict this flourishing industry and the international trade it supported led to the American Revolution.

Morison's (1941) argument that the constriction of maritime trade led to Shays' Rebellion indicates linkages much tighter and more dynamic than those that prevailed in Poland and Germany between logging and the expansion of a dynamic yeomanry. The timber trade stimulated the wheat trade in Poland and Germany and led to changes in agrarian structure, including a major subordination of rural labor in the "second serfdom" (cf. Kriedte 1983; Wallerstein 1982), but it could not stimulate industry there. Dutch merchants could supply imported manufactures more cheaply than they could be produced locally. New England loggers, farmers, and fishermen, in contrast, were all part of a population that could turn as well to manufacture and produce commodities more cheaply than they could be imported.

Perhaps even more important, though, was that the various bulk goods that dominated both Dutch and American trade did not require huge capital outlays or the development of hugely expensive naval ships. The shipment of timber and other bulk goods created most of the value in the commodity itself; Albion (1926) calculated that a log in the Polish woods was priced at only 5 percent of its eventual British price and that up to 75 percent of this difference was made up in shipping costs. In the high-value, low-volume Asian trade, the cargo itself might be worth ten times the value of the boat. Such trade was only accessible to highly capitalized merchants and needed the protection of an expensive navy.

Much of what the New England traders exported, in contrast, were high-bulk staples, which did not require the huge capital per cargo or the naval protection that preciosities and finished goods did. Thus there were fewer capital barriers to entry; most of the value was added by transport. Returns on wood and cotton cargos or on shipments of ice and granite might return far less profit, but once the boat itself was paid for and provisioned, these lower-value cargos cost much less to purchase and were therefore accessible to smaller capitals. Even when New England ships entered the China trade, they did so by taking sea otter from the northwest coast, achieving directly a huge transport-based creation of value that allowed exchange for preciosities (Morison 1941).

Morison (1941) calls shipbuilding New England's leading industry. He shows that until the Revolutionary War and the War of 1812, trade was carried on from many New England ports but that the wars and the raids on some ports did allow Boston to centralize that business. Prior to that, though, there were many different boatyards and owners. These yards both consumed and traded timber, and the timber trade flourished. After whale oil, timber was the major New England export (Davis 1962). Portland's supply of timber to Cuba's sugar mills, for instance, was phenomenal.

The 200-odd distilleries of rum in Boston as well as the textile mills and shoe factories that emerged in the river valleys were directly dependent on shipping, as was, of course, the fishing trade. The shorter coastal trade, however, also handled amazingly bulky materials, including monumental granite blocks for public buildings in coastal cities and ice to New Orleans and points between.

The extraordinary concentration of the whaling industry around New England, even though the hunting territory was global, clearly also drove and was supported by shipbuilding. The fur trade on the northwest coast involved multiple international connections around Cape Horn with stops in Valparaiso, negotiations with the Russians in Alaska, trade with Spanish California, and negotiations with the Dutch and the Chinese. By the time the gold rushes in California and Australia stimulated both migration and agricultural exports to those areas, New England shippers were well positioned to take advantage of the expanded demand for carriage. New England loggers, in turn, benefited from the expanded demand for ships.

The small size of New England ships reduced capital risk. They could be put together quickly at low cost because of the cheap timber. They did not compete for raw materials with other industries such as metallurgy because of the cheap coastal trade between different regions, though shippers did oppose the tariffs that the fledgling textile industries required. Their other tremendous advantage was the huge migration to America; there was a great deal of money to be made returning from a high-volume cargo delivery to Europe with a human cargo. In the early days, the distribution of boatyards in multiple harbors on a highly indented coast had tremendous economic benefits and tremendous spread effects.

In addition to shipbuilding skills and labor, merchant capital was drawn to New England by the cheap timber and trade opportunities. As in Holland, shipbuilding stimulated ancillary industries such as iron working and sail-making as well as food preparation for crew provisions. The peculiar to-pographical advantage of numerous short, rapidly falling rivers that so favored shipbuilding also provided cheap water power for textile and shoe factories, as well as sawmills. Textiles and shoes increasingly served both to link with trade and to absorb labor during the periodic declines in fisheries.

Capital and Labor: From the Atlantic Coast to the Continental Interior

Two wars in close sequence concentrated trade in Boston and New York and ended the decentralizing advantages and industrial linkages of multiple boat-yards and ports. British blockades and attacks ruined the smaller ports. The diffusion of early shipbuilding, fishing, and industry to multiple small ports fed by rivers that drained ecologically diverse valleys, though, had clearly contributed to local accumulations of capital and to the development of a national society. The restriction of American shipping and trading rights to British ports during and after these wars encouraged New England capital to increase internal trade. Coastal shipping expanded; as the economy spread inland, transport and trade followed. Restricted access to Europe also drove shippers to trade with Latin America and with China, creating knowledge of and presence in the Pacific Ocean.

America was too important a source of raw materials for Britain to persist in punitive restrictions on trade. As American shippers turned their attention to the Pacific, British mechanization and expansion of the textile industry created a huge European demand for American cotton. Pressure from mill owners quickly relaxed the restrictions on American shipping. The renewed Atlantic economy thus combined with the opening Pacific economy to create a flourishing market for transport on both coasts and across both oceans from the 1830s to the 1860s.

Independence from and eventual peace with Britain were followed by mili-tary and diplomatic negotiations with the various European powers—British,

French, and Spanish—that claimed parts of the American continent's interior. The European powers were too involved in contests with each other to sustain ongoing struggles over distant American territory, so the United States prevailed fairly easily in most cases and negotiated effective truces after brief, low-intensity confrontations in the rest. This enabled a rapid, relatively uncontested expansion of U.S. sovereignty over huge, fertile, highly transitable space just as Britain was improving iron, coal, and steam technologies sufficiently to make their use in mass production and transport commercially viable. The Americans vastly expanded the economies of scale and speed that the new mineral technologies made possible to accelerate extraction of the sylvan, agricultural, and mineral resources that abounded in their vast, sparsely populated territory.

Individual entrepreneurs and small groups of investors in coastal cities had accumulated significant capital in logging, shipbuilding, shipping, and fishing. The western frontier presented opportunities that they could seize only by investing in infrastructural projects whose costs exceeded their individual capitals. Municipal associations of merchants pooled their capital and state governments sold bonds to finance highways and canals from their various coastal city-states to the rapidly broadening western hinterlands.

Vance (1990) and Meinig (1993) both characterize the competition between cities for access to opening hinterlands as a reflection of the original "mistake" of locating them along the Atlantic coast. In fact, when these cities were founded, their locations responded to the prevailing technologies and markets and to the available raw materials. Subsequent campaigns to build highways and canals to the economically and demographically expanding west responded to changes in production technologies that required new and expanded types of raw materials, urban labor forces that required more and cheaper food, and transport technologies that could move large volumes more cheaply. These cities had accumulated capital adequate to their need for new and expanded transport systems precisely because they had been in the best locations to take advantage of earlier markets and transport technologies.

The local impulse to link established cities to new and expanded hinterlands paralleled and fed into national efforts to gain access to more land and natural resources and to secure access to markets. The interaction of local and national expansive impulses coincided with and made possible the citizenry's ambitions for land and employment. Public-private collaboration in the acquisition of territory, in the development of new transport technologies, and in the construction of transport systems converged with the evolving collective conscience of the nation. This national conscience linked territorial expansion, egalitarian access to land and matter, productive labor, and broad commerce to the formation of a vigorous and "virtuous" republic.

The international and domestic policies and pronouncements of early U.S. presidents reflected and articulated the national population's keen interest in and sensitivity to the economic and demographic opportunities of America's interior space and to the political benefits that its incorporation and settlement by commercially oriented yeoman farmers would bring. Their struggles against Britain and the structure they subsequently created for American government were highly influenced by the egalitarian, communitarian philosophy of John Locke.

Locke had formulated an early version of the labor theory of value, based on the argument that the deer was nobody's property until the Native American shot an arrow into it, thereby incorporating his labor into the deer. While this vision validates the Native American's labor, it simultaneously affirms the right of anyone else to take a deer, or any other natural product, into which no labor has yet been incorporated. In other words, it allows value, property, and individual and community rights only to social, and not to natural, products, and it affirms that labor confirms rights in property while it benefits the community. As such, it justifies appropriation of matter and space that does not directly and immediately serve social production. It thus served to legitimate colonial expansion as generating labor value and creating property rights from what would otherwise remain the "free gifts of nature" only sporadically and partially exploited. From this perspective, imperialism worked to the advantage, enlightenment, and advancement of colonized peoples, even if they had to be violently subordinated to accept it.

These British imperial principles were incorporated into emerging and progressively more aggressively ethnocentric U.S. state policies toward indigenous populations. Jefferson, as president, clearly stated that the extensive breeding and grazing areas required for the reproduction of wild plants and animals that sustained indigenous populations or that were claimed but not exploited by distant European monarchies were to be divided into individually owned plots of land of a size suited to family labor for commercial profit—640 acres per family with additional acreage for each child (Ellis 1997).

Early American national plans and projects thus proposed and justified incorporation of specific territories in terms of their economic potential—

extractive and agricultural—and their particular natural avenues of transport to commercial markets. George Washington's early career as land surveyor and military officer in British colonial campaigns to secure the upper Ohio River Valley had sharpened his geographical knowledge and imagination and turned it toward a strong belief that the young nation's future lay west of the Appalachians. Thomas Jefferson, in his prolific writings as well as in the territorial policies and programs he instituted domestically and internationally, combined a strong belief in the civic benefits of a vigorous independent yeomanry with a keen appreciation of the geography of transport and security. Like Adam Smith, with whom he maintained a long correspondence, Jefferson was deeply influenced by the Physiocratic theories of Quesnay and Du Pont de Nemours. Quesnay believed that all authentic wealth originated in agriculture. He argued that good governance—rooted in and financed by the free commerce of land and its products—could achieve civic virtue and prosperity only if it promoted agricultural production and guaranteed it fair prices.

Jefferson synthesized Locke's principles of labor value and of the social contract with Quesnay's ideals of land-based prosperity and civic virtue into an idyllic agrarian democracy rooted in a deep understanding and imagination of how American geography—including that of territory still unexplored and under foreign control—would affect production, transport, and commerce. He was passionately committed to securing an outlet to the Gulf of Mexico. He wanted to assure safe passage around Cuba in order to connect the Mississippi Delta to European markets. His successful completion of the Louisiana Purchase achieved the outlet to the Gulf, opened the continental interior to U.S. settlement and control, and eventually led to the displacement of Spain from Florida and from the southwest. It catalyzed as well the definitive push northward of the United States against Britain's territorial claims in the midwest and northwest.

Jefferson's sponsorship of Lewis and Clark was driven by even more visionary imaginings of production, transport, and exchange based on his long-time dream of finding a continuous water route between the Mississippi and the Pacific. Commerce with Asia had gained importance as renewed tensions with Britain restricted U.S. commerce with Europe. Chinese goods were generally high value and low volume; China sent goods of greater value to Europe than it took in exchange and so absorbed a surplus of Europe's gold and silver. The American wilderness offered abundant supplies of fine fur. The Chinese particularly prized sea otter and seal pelts from the northwest coast.

Jefferson dreamed of enhancing this high-value/low-volume trade by opening the far west to access inland from the Pacific along the Columbia River. He had interpreted the extremely sketchy maps of the northwest as possibly indicating a water connection between the Missouri and the Columbia Rivers. Lewis and Clark encountered instead "tremendous mountains"—the Rockies —but they did claim the territory all the way to the Columbia's outlet to the Pacific for the United States, enabling John Jacob Astor to set up a far western trading center intended as an entrepôt between the Pacific trade and overland routes back to the Atlantic coast.

Jefferson's vision of a republic of yeomen expanding agriculture and commerce into the spaces opened by the trade of fur for imported preciosities did not leave room for indigenous inhabitants or for their reliance on self-reproducing, broadly dispersed herds of animals. Jefferson at first proposed to assimilate the Native Americans as agriculturalists, distributing the vast parts of their territory that would thus remain unused for European occupation and use. Soon after becoming president, though, he announced plans to relocate the Native Americans who remained east of the Mississippi to the western part of the new territories the United States had acquired and to use those tribes to supply the fur trade until the agricultural frontier reached those territories as well.

To this end his government established "factories" that exchanged manufactures from New England for furs. The private fur-trading companies—especially Astor's—were far more rapacious—and tended more to aggravate relations with the Native Americans—than the government "factories" did, but they were more profitable. The federal government depended on these traders to extend and defend its still-tenuous claims to a sparsely occupied frontier and therefore did little to stop them or restrain their provocations. As in the Amazon, hunting for profit decimated the animal population and displaced the human population that depended on it. Violence reinforced the loss of food as negotiated displacement of the Native Americans steadily gave way to military campaigns to assure productive and commercial space for U.S. settlers and capital.

Though America's westward movement was temporarily interrupted by war with England less than ten years after the Louisiana Purchase, Jefferson's Lockean ideals about labor value and his Physiocratic principles about agrarian life and work, energized by his geopolitical imagination, ethnocentrism, diplomatic skills, and enterprise, tremendously influenced the course of U.S. territorial and economic expansion. George Washington and many other leaders of the revolutionary and constitutional movements believed passionately that the energies of America's immigrant population, freed from the shackles of European political and economic class distinctions and restrictions, would shape the natural wealth, size, and topography of the continent into a nation of unmatched prosperity and enlightenment (Ellis 2000). Jefferson mobilized population and resources to realize this belief. He eloquently and effectively articulated the geographical conscience of the nation around its historically unique configuration of space, matter, technology, population, and culture.

Jefferson's presidency coincided with the opening up of new Atlantic trading opportunities for American merchants as a result of the Napoleonic Wars, the growing New England industrialization, and the significant migration from Philadelphia and Baltimore into the Ohio Valley. His articulation of widely held ideals, combined with his aggressively expansionist policies, played powerfully into the formation of a moral universe that assured great rights and protections to the commercially oriented social production of its members but few to other production systems, whether they be social or natural. It accentuated creative intertwining and circular reinforcements between imaginings of the continent's economic geography and the technical and social-organizational solutions to its challenges and opportunities. His visions of the material configurations of a still largely unexplored surrounding space reinforced the yearnings of a population willing to occupy and transform that territory to enhance its political and economic power.

The United States reiterated on a far grander scale the technological, infrastructural, political, and economic innovations that the Dutch invented to build their dikes, canals, dams, windmills, and polders. The institutions linking labor and capital from the separate locations that would benefit from such projects required the creation of authorities commensurate with the extra-local scope of the territories that these projects would affect. The expanded moral community and evolving state capacity functioned first to reclaim flooded lands. Once the institutional and technical bases for state capacity had been established and were functioning effectively, state authority and capacity increased sufficiently to coordinate across broader space the collective deployment of multiple local municipalities to finance and construct the instruments of transport, exchange, and production that eventually made Holland the world's richest nation.

In the United States, the challenges and opportunities that the population and the evolving national state saw in a vastly larger territory catalyzed a similar interaction of spatial, material, and technological visions. American agricultural, extractive, and industrial opportunities attracted successive waves of European migrants anxious to turn their agricultural, mining, and industrial skills to commercial profit. Even so, the American economy expanded faster than the labor force did, creating strong incentives to invent and implement labor-saving technologies for all branches of production. The ethnic and cultural diversity of the rapidly growing population, the land-hunger and acceptance of risk and of hard work, and the commercial ambitions that attracted a large portion of the immigrants facilitated the formation of a national culture that favored territorial expansion and technological innovation (Parker 1991).

These same characteristics, enhanced by national policies favoring small agricultural properties, promoted social principles of individual liberty within a sense of responsibility to (and protection by) community, a strong belief in individual rights to property, and an inclination to insist on a fair return for the products of the individual's land and labor. These cultural and political attributes created a disposition to contest capital's power to abuse or exploit labor while strongly supporting capital's territorial expansion. They also created a clear sense that the state was beholden both to labor and to capital—and that state promotion of territorial expansion would benefit both. The emerging collective conscience included a strong sense of geographical or environmental potential with an enthusiasm for its realization through expanded transport infrastructure and more efficient transport technologies (Parker 1991; Meinig 1993).

How Matter, Space, and Technology Formed the American Polity

As we saw in the Amazon, newly opened distant territories are typically incorporated into the world-system through a sequence of raw materials exports, starting with very high value-to-volume goods and progressing to goods of higher volume-to-value as transport technologies and infrastructures improve. The greater the distance from established lines of trade and communication to the core, the more likely that this incorporation will reduce indigenous populations by appropriating and depleting the material bases of their subsistence by introducing new diseases and by military violence.

In the Amazon, the triple scourges of resource depletion, exotic disease, and violence left the rural areas so sparsely populated that when a higher volume-to-value export—rubber—required a significant work force, labor had to be imported from the Brazilian Northeast. The Amazon's humid tropical ecology and distance from markets discouraged European agriculturalists from replacing indigenous inhabitants, so the demographic vacuum persisted.

In contrast, North America's temperate climate and soil types made European technologies and practices not only commercially viable but actually more productive. Land and raw materials were more abundant and much cheaper, so labor was more productive. Scarcer in America, it was also better paid.

Migration, though, was costly, dangerous, and laborious. Migrants to America tended to be those whose productive skills and commercial ambitions offered them sufficient potential income to offset the travails of the journey. Nonetheless, successive waves of migrants arrived in great numbers from different European nations.

Local and national political and economic culture evolved around the interaction of this diverse, skilled, and commercially ambitious population with the material and spatial opportunities and challenges the new continent offered. Migrants brought skills, technologies, social and economic attitudes, belief systems, and purposes that had evolved in more densely populated, more commercially integrated, and smaller, more meager environments than North America. They adapted all of these social attributes to the larger spaces, richer material bases, and less rigid, more egalitarian social and political organization they encountered in America.

In the process, they learned new ways of handling matter and space and developed ambitions for more of both. The nascent collective conscience, or national political and economic culture, was formed in frequent and intimate experience of adapting, innovating, and improvising technologies and community organization to overcome the challenges and to exploit the opportunities that an environment they were still exploring presented to them. Many speculated about how transport through that environment might give them access to its resources and about how they would have to organize collectively to ensure their safety (Meinig 1993).

Jefferson's geographical genius, agricultural ideals, and nationalistic aspirations appealed to and inspired these imaginings, while giving them articulate, eloquent, and strategically astute form (Meinig 1993; Ellis 1997). The national state gained in authority and legitimacy as it appropriated the territory and provided the transport means that both capital and labor needed to realize these imaginings, first through support for canals, highways, locks, and dams and then on a much grander scale for railroads. Though initially impecunious, the national state was able to finance its transport projects, and in the process increase its own power, by sales and grants of the lands it and its citizens had appropriated.

The realization of these imaginings was as devastating to the existing ecological and social systems in the United States as the European conquest and exploitation of the Amazon was. In America, though, a new and economically vibrant population replaced the indigenous one it expelled, and it established new forms of production, replacing the depleted beaver, buffalo, sea otter, and forests with wheat fields and cattle pastures. Jefferson's expansionist policies had simply expedited the first steps in this process, opening the Mississippi to agricultural and timber exports and the far west to the far less voluminous trade in precious furs.

The state legitimated the large corporations that dominated the fur trade as necessary precursors to settlement of the west by yeoman farmers, but the frontier they were opening was distant enough that the corporations could expand and enforce their areas of trade control through violent force, with little or no effective state opposition. Large monopolies over land, resources, and trade rather than communities of yeoman farmers dominated the incorporation of the continental west into the nation's territory.

Jefferson's agrarian democratic ideals of settling yeoman farmers on land adequate to their own family labor was embedded in a highly expansionist disregard for the property and political rights of any groups, indigenous or European, that did not assimilate to his vision of community. The prosperity and safety of these agrarian communities required land, and Jefferson's willingness to use large and rapacious fur-trading companies, like Astor's, to extend the nation's territorial claims set the stage for the dominance of large capitals to prevail in expansion beyond the Mississippi—lobbying for huge grants of public land to support their investments in canals, railroads, and mines; colluding to gain monopoly control of essential transport, storage, and trade facilities for agricultural and extractive goods; and suborning the state to ignore their violent repression of labor. Through long and often violent struggle, labor was able to check some of capital's growing power domestically, but we will see in later sections how these expansionist forces enshrined in Jefferson's early nineteenth-century idealism were by the end of that century and through the next driving imperialist endeavors abroad.

The expanded agricultural and extractive commerce on the Mississippi had by 1825 inspired rapid adaptation of British coal and steel technologies to American distances and volumes. The Mississippi, and then the Great Lakes, became the nursery of steamboat innovations. Cheap, fast lake and river transport stimulated intensified extraction and cultivation of higher volume-to-value goods on the surrounding plains and tributary valleys. Water transport eventually merged with an expanding rail network that articulated huge deposits of metallurgical coal with huge deposits of high-quality iron, delivering both for mass industrial production all along the southern shores of the Great Lakes.

In ways strongly reminiscent of the interaction of Dutch cultural formation with the challenges and opportunities of their particular environment, the culture that drove these technological and economic innovations was formed in the crucible of immigrant populations' innovative responses to local matter and space as they discovered and exploited it. If charisma manifests the ability to express insights about society and nature that simultaneously articulate the deep yearnings of a population and mobilize that population to realize those yearnings, Jefferson provided the earliest charismatic expression of this expansive culture, galvanizing a newly forming nation around a rapidly forming— and increasingly capable and powerful—state at the beginning of the nineteenth century. By the end of the century, the new iron and steel technologies had enabled the nation to incorporate fully the lands opened by Jefferson's transport and agricultural geographic aspirations.

How Iron, Coal, and a Canal Finally Gave Britain the Carrying Edge

Large ships were increasingly made of iron and steel rather than of wood. Britain, though rapidly losing ground to Germany and the United States, still dominated world iron and steel production. British metallurgical skills and supremacy in steel production gave her boatyards a tremendous comparative advantage in the design and fabrication of steam engines and the metal hulls they powered. As Cafruny (1987) put it, the British steamship ended "the potent American challenge," relegating "the United States to the status of a minor maritime nation" (49). Building the Suez Canal required unprecedented mechanical and hydrological technologies; complex negotiations with Ottoman and French governments over rights of access to, property in, and passage through Egyptian territory over which all three powers claimed rights; and financial agencies capable of funding huge investments in a project in an uncertain and volatile region far from national territory.

The commercial viability of the entire project depended on the continuing technological and infrastructural advances in steamship navigation and in the railroads that brought grains and metals from the huge Asian interior to the coasts of the Indian Ocean. Iron hulls and coal power had enabled British gunboats to fight their way upriver into China's vast interior as early as the 1840s, opening that country to the triangular trade between Indian opium, Chinese tea, and British capital, transport, and beverage markets. Shipping, though, remained dependent on British state subsidies until the Suez Canal was completed, cheapening transport enough to allow the export of higher-volume, lower-value goods like rice as well as the growing commerce in tea. As Headrick (1981) writes: "In the Indian Ocean, the introduction of unsubsidized shipping lines coincides with three technological advances . . . : the compound engine, the Suez Canal, and the submarine cable. The cable . . . allowed the headquarters of the shipping companies to maintain contact with their ships, and shippers to coordinate their shipments . . . with the needs of their customers" (172).

These strategic victories and technical accomplishments were reinforced by Britain's sponsorship of Chile's War of the Pacific against Bolivia and Peru in the 1870s. Chile's victory wrested the mineral-rich Atacama plateau from its neighbors to the north and gave British capital the open access it had sought to the nitrates there, enabling the expanded investments required for that raw material to revolutionize both agriculture and warfare by providing new bases for fertilizers, as well as for explosives and bullets. Britain's iron-coal-steam technologies were critical both to building the warships it gave Chile for this adventure and also for the mining, processing, and shipping of nitrates back to Europe.

It was Britain's booming industrial economy that created the demand for raw materials, and it was the close collaboration of the state, finance, and industry to assure cheap and stable access to them that drove these imperialist ventures. Quiroz Norris's (1983) close analysis of British commercial and financial investments in South America during the guano and nitrate booms directly supports our two arguments that commerce, finance, and politics follow into areas first incorporated into the world-system by the quest for raw materials, and that this occurs at the height of investment in the trade-dominant nation rather than when capital has become over-accumulated there.

Sailboats remained more efficient carriers on a cost per ton-mile basis than steamships until the 1880s, but they had to be towed through the Suez Canal (Fletcher 1958). The costs of transporting coal, however, and the tremendous advantages available in the India trade drove the adoption of a series of innovations in motor design that reduced coal consumption per horsepower-hour from the eight to ten pounds of the 1830s single-cylinder motors to the two pounds of the 1860s compound engines. The even more efficient high-pressure triple expansion engine of the 1880s finally doomed the sailing ships. The boom in steamboat construction significantly stimulated machine tooling and engine-building technologies, as well as stimulating the sophistication of specialized production units and their reincorporation into a large, complex production process (Hobsbawm 1968).

The success of the steamboat rested on the earlier development of another metal-based, coal-driven bulk transport vehicle and its infrastructure. The massive development of railroads earlier in the century had revolutionized industry and finance. Hobsbawm (1968) credits railway construction in Britain during the 1840s with the investment of over two hundred million pounds, direct employment of 200,000, and the doubling of iron output, accounting in 1845–47 for "perhaps forty percent of the country's entire domestic production, settling down thereafter to a steady fifteen percent of its output" (114). Hobsbawm also credits the railways for stimulating engineering training and innovation and with direct stimulus to the capital goods markets that moved the British economy away from its precarious and technologically simpler dependence on textiles.

Hobsbawm (1968) maintains that the steamship was "not capable of competing with the increasingly efficient sailing ship until the revolutionary transformation of the capital-goods base of the industrial economy which the railroad era inaugurated" (114). The combination of railroads and steamships with the vastly expanded demand for the wide variety and huge volume of inputs they created (and whose cheap transport they were able to provide) set the stage for Britain's short-lived industrial dominance and her much longer dominance of world trade and finance.

The expansion of world trade to include not just the traditional base metals but also the metals required for special alloys in machine tools, together with the vastly expanded markets for grains and meat, boosted the need for railroads around the world. Britain constructed boats and trains and exported them and the coal to drive them around the world. Employment and wages soared, new financial institutions were created, and vast amounts of capital were invested abroad between 1870 and 1914, with 41 percent going directly to railroads, another 5 percent to ports and ships, and 30 percent to colonial government, of which a substantial share went to transport infrastructure (Hobsbawm 1968; Latham 1978: 54–55).

In addition to restructuring, deepening, and accelerating Britain's industrial economy, her dominance of railroads, steamships, and the Suez Canal allowed her to control a vastly more complex and tightly integrated world transport system than had earlier been technically possible or economically advantageous. Simultaneously, the transport revolution stimulated an enormous increase in the transmaritime shipment of industrial raw materials. Rail, port, and shipping systems all repaid integrated organization around the set of technologies that Britain controlled. As new world market opportunities for raw materials emerged, national governments could be induced, and colonial governments directed, to invest vast sums in overland transport infrastructure and in ports.

Technologies built on iron, coal, and steam enabled far more rapid increases in economies of scale than had been possible with wood. The technological innovations that made the new scales of production and of internal transport possible often involved revolutionary changes in tools and process, but between each revolutionary advance were multiple incremental technical improvements. These enabled most of the investments required to realize these expanded economies of scale to grow at a rate proportional to these incremental changes. For the most part, the institutions that financed them and the organizations that managed them could expand at similarly incremental rhythms. Because transoceanic shipping required technologies capable of traveling long distances between taking on more fuel and water, though, moving steam engine technologies from overland and coastal to maritime transport necessarily involved much larger leaps of scale than the incremental expansions possible in internal transport or factory systems.

The spatial challenges of transoceanic technological innovations fomented major generative responses in British politics and finance. In addition to having developed the world's most sophisticated metallurgy, Britain possessed as well its most powerful and sophisticated financial system and a long experience of imperial and diplomatic negotiations. These national attributes enabled Britain to innovate and expand financial instruments and institutions—domestically and abroad—at the rate required to take and keep the lead in the parallel development of iron hulls and more powerful motors for oceangoing vessels. The new financial institutions became especially adept at creating and managing the new instruments of credit and debt—bonds, mortgages, and corporate stock—that served to distribute the construction and infrastructural costs to raw materials supplying nations.

Nineteenth-century British finance thus invented the first iteration of a process—globalization at peripheral expense—whose latest vastly expanded iteration we saw in the debt-based financing for Carajás. Iron-coal-steam–based economies of scale raised the cost and increased the speed of establishing new technologies worldwide; core financial innovations accompanied changes in technology and changes in national ideologies and the world-system by steadily expanding the scope, scale, and speed of financial instruments adequate to intervene in core nation budgets and policies.

Britain reinforced her growing commercial advantage by establishing chains of fueling stations at appropriate intervals along her widely dispersed trade and administrative routes and by stimulating massive investments in the infrastructure and environmental modifications required for the integration of overland railways with transmaritime steam-powered transport systems in huge and often contested spaces. Some of these investments were arranged in countries under direct colonial control, others in Latin American and Asian nations that were putatively independent, and some in territories that were either struggling for independence or subject to competitive core claims to control. In most cases, though, Britain managed to impose significant debts on the state systems that formally governed these territories.

By thus combining domestic technological and financial innovations with external geopolitical strategies and powers, Britain was finally able to dominate the world maritime industry. The state, the navy, and national capitalist classes had aspired to this achievement for over two centuries, but could not succeed until maritime transport technologies were based on minerals rather than on wood. Britain's territorial control of abundant, high-quality coal that was easy to mine and to transport provided the materio-spatial advantage that primary reliance on naval force could not. This advantage finally enabled the state, the navy, national firms, sectors, and dominant classes to marshal the different technical, financial, diplomatic, and organizational skills and capacities needed to combine environmental modifications on an international scale with building a massive new fleet. They could then impose a regime of free trade that enhanced their growing material trade advantage.

Asian and Latin American railroads were critically important to this undertaking, but the Suez Canal provided its essential base. The construction of the Suez Canal in 1869 and the canal's contribution to the rapid growth of the India trade, which in turn accelerated the technological improvements of steamship engines and hulls that the canal enabled to participate profitably in that trade, assured that Britain conclusively became the world-dominant carrying nation. According to Cafruny (1987), coal had replaced wood as the bulkiest input in industrial production: "Britain had vast reserves of coal located near navigable rivers and coastal ports, and great experience in mining it. The coal trades were the single most important basis of British maritime expansion, providing ready cargos and cheap propulsion for British shipowners" (49).

U.S. energies, transport investments, and raw materials procurement continued turning inward during this time. U.S. coastal trade volumes had surpassed transmaritime volumes in 1831. Two decades later, the gold rush to California and the further settlement of the northwest coast attracted a great deal of Boston shipping. The growing supplies of western meat and minerals intensified the long-held dream of overland transport from East to West. In 1869, the same year the Suez Canal was completed, a complex collaboration between local and national governments and regional and national financiers completed the rail line linking the Atlantic and Pacific. By 1890, British shipping capacity was twice that of the United States. Its supremacy in steamship transport, which soon dominated international trade, was even more secure; according to Cafruny (1987), by 1880 Britain had over twice as much steamship capacity as all other countries combined (51).

Britain's only period of dominance in world trade, the period of free trade based on the pound sterling and the period of her industrial dominance (Hobsbawm 1968; McMichael 1984), corresponds to the period when Britain dominated the world's metal-based transport, that is, 1870 to 1914. The end of the sailing boat era ended her dependence on military and colonial strategies to overcome the high cost of raw materials to build ships. The raw materials advantages she enjoyed in building transport vehicles and infrastructure stimulated employment and new technologies, made her industrial and capital goods exports highly competitive, and allowed her not just to export raw materials and heavy industrial goods in her own vessels but also to bring the ships of other European nations to her ports in order to find bulky outbound cargos in coal (Fletcher 1958). Britain was therefore extremely well positioned to assure cheap access to the raw materials to make new metals for the increasingly sophisticated metallurgy that was emerging from the transport revolution.

In this period of British history, we can see a repetition on a far larger scale of the earlier Dutch experience. Both national economies rose through a self-reinforcing spiral that started with privileged access to the raw materials critical to the most efficient transport technologies of the era. Both nations used this access to expand and improve these transport technologies and then used the new technologies to cheapen and expand their access to the raw materials needed to build and operate the new transport systems. Cheaper, more abundant raw materials facilitated higher throughput in all industrial sectors— including shipbuilding. The higher throughput allowed experimentation with and the adoption of new technologies and new organizational forms for both capital and labor, particularly in the handling and transformation of large volumes of bulky matter. It also sustained complex reorganizations of the increasingly distant peripheral economies that provided cheap and stable access to (and often internal transport of) critical raw materials.

This further cheapened raw materials extraction and transport from the periphery while heightening labor productivity and rates of profit in the core, particularly in the transport sector. Cheap national industrial products became an important source of export revenues. Expanding trade increased the vitality and power of the national financial sector, whose institutional forms restructured world financial relations to the advantage of the transport-dominating economy.

Because British dominance of wind-driven world trade had depended on coercive imperial policies and on the trade-restricting Corn Laws and Navigation Acts, it had not generated dynamic and self-sustained economic and political development comparable to that generated by the Dutch transformation of naturally occurring spatial and material advantages into trade supremacy. When new technologies made steam power and metal machines preeminent in world production and trade, Britain's ample and conveniently located coal deposits finally provided her with natural advantages comparable to those that had generated Dutch financial, commercial, and political innovations and world leadership and launched her on a similarly progressive spiral.

Simultaneously advancing American efficiencies in supplying industrial materials enabled the British to spin this spiral out on far broader and richer spaces than had been available to the Dutch (cf. Wallerstein 1982).

The impact of the British raw materials advantage on domestic industry was short-lived, however. Indeed, the shipyards may simply have forestalled the decline from world dominance that the centuries of metallurgical development produced in the form of the railroad boom. Britain dominated world production of ships until the extraordinary growth of U.S. shipbuilding during World War I, but maritime transport was no longer the sole industrial product for transporting bulky raw materials. Germany and the United States were combining internal water and rail to supply their own rapid industrialization and could capture the advantages of rapid growth to introduce new techniques that were relatively more costly for the British physical plants already in existence.

Ironically, the British steel industry failed to adopt its own innovations as rapidly as its German and U.S. competitors, in large part because the efficiencies of British maritime transport made it cheaper in the short run to import iron and steel than to retrofit existing plants (Hobsbawm 1968). British metallurgy led the world in the 1840s and was thus able to support and benefit from the railroad boom. Because it came later, railroad building in Germany and the United States could take advantage of technical developments in metal processing that occurred at the peak of railway construction in Britain (Landes 1969; Hugill 1994).

The greater national spaces in Germany and particularly the United States created an expanded market for iron and then steel rails. The larger markets supported new smelters with expanded capacity that could profitably incorporate new technologies that both lowered the cost of steel *and* improved its quality. German and American increases in smelter scale contributed to the rapid decline of the British share of world steel production after 1860. In both countries, territorial size and resource endowments enabled internal rather than external access to all of the raw materials needed for steel. In both countries, these raw materials constituted the most voluminous cargos, increasing each economy's ton-mile requirements for transport. In both countries, existing waterways were improved to enable heavier, faster shipments, and railroads developed quickly, first as a complement to internal water travel and then increasingly as a more rapid substitute. The total miles of rail built in each of these countries soon surpassed Britain's total.

In Britain, the development of the internal rail system preceded, and then supported, the development of an external steamship system. In the United States, wood-burning, wooden hulled steamboats for coastal and river transport imposed less capital risk than railroads. America therefore reversed the British sequence; steamboats preceded and supported the railroad boom. The first railroads of the 1830s and 1840s were primarily short feeder tracks into canals and river systems or dedicated lines from coal fields to the nearest coastal city. In America, water transport remained supreme until the Civil War.

In America, the first commercially successful steamboats operated on the Hudson River, the Chesapeake Bay, and Long Island Sound. They soon moved to other parts of the Atlantic coast at about the same time that the Louisiana Purchase opened the Mississippi, its tributary valleys, and its rich surrounding plains to U.S. occupation and commerce. Fulton and his associates had seen their Hudson River enterprise as simply the precursor to their expansion to the Mississippi and to the rapidly growing economies along the Ohio River. Numerous other intrepid entrepreneurs shared this vision of profits from the provision of transport to Mississippi commerce, and by 1812 the first steamboat reached New Orleans from Pittsburgh, where it had been constructed.

As Jefferson had anticipated, the expanded U.S. territory and its easy outlets to international and East Coast markets intensified the already vigorous agricultural settlement and production of wheat, pork, and cotton. Steam transport also intensified extractive flows down the Mississippi. These were no longer limited to the high-value, low-volume beaver furs that had opened the upper midwest to commerce two centuries earlier; it now included growing volumes of logs and lumber, as well as heavy cargos of lead and zinc from Illinois and Wisconsin. When steamboats reached the upper Missouri, they brought back huge bales of buffalo skins. Steam enabled upriver transport of manufactured goods and tools from the Atlantic coast, cheapening the costs of subsistence, enhancing labor's productivity, and so encouraging more settlement.

From Artisanal Iron Plantations to Mass-Producing Steel Mills

The long, spectacular rise of the American steel industry and its profound and sustained effects on U.S. political economy and ecology reiterate dramatically the complex and intimate interweaving of increased volumes and types of matter, extended space, more powerful technology, larger financial institutions, an increasingly commercial political and economic culture, and a stronger, more capable state—all intensifying, extending, and reinforcing each other over time—that ushered in a new and expanded systemic cycle of accumulation. The size and rich resource base of the territory the new nation claimed as unoccupied; the absence of traditional forms of property, authority, and economy; and the coincidence of expansion into this territory with the rapid development of coal and iron technologies in Europe magnified and accelerated the self-reinforcing effects of what rapidly became self-sustaining generative sectors. The U.S. steel industry repeated in expanded and accelerated form the development sequence of the generative sectors of other rising nations.

Abundant forests enabled U.S. iron producers to continue their reliance on charcoal fuel long after Britain became dependent on coal, but the Pennsylvania forests that supported much of the U.S. industry were depleting. Anthracite coal was abundant there and was extensively used to heat urban residences and factories, but it did not easily provide the high heats needed for smelting. The Ohio Valley, in contrast, still had extensive forests and large deposits of iron ore. The already substantial flow of migrants into the Ohio Valley increased when the Louisiana Purchase opened the Mississippi to bulk transport down the river for transshipment from New Orleans to Europe and to the Atlantic coast. The initiation of steam transport, and its ability to move bulky cargos up- as well as downriver, hugely expanded its iron industry. All through the Ohio Valley, oak forests around deposits of iron close to rivers supported iron "plantations." Their spatially dispersed production provided raw materials for the tool and machine industries—including boatyards for steamboat engine and hull construction and repair—emerging in the small towns along the Mississippi and Ohio Rivers.

Steamboat transport was inaugurated on the Great Lakes shortly after the War of 1812. Niagara Falls and the rapids on the St. Lawrence River impeded lake transport of bulky goods for international commerce until a series of major infrastructural projects on both sides of the Canada-U.S. border—including the Welland Canal and, most notably, the Erie Canal—opened water traffic to the Atlantic coast. The Erie Canal stimulated huge flows of timber, wheat, and flour from the upper midwest and cheapened immigration for the laborers, loggers, and farmers who would expand that flow.

By 1840, loggers discovered major deposits of both iron and copper in northern Michigan. Rapid technical advances and infrastructural extensions in railroads, steamboats, and agricultural implements and machines, together with the increased use of metal in constructing America's rapidly growing urban-industrial and agricultural-commercial centers, promised vibrant demand for both minerals. New England capitalists, supported by land grants from the state of Michigan, combined their separate capitals to finance locks and dams around the Sault Ste. Marie falls, enabling the cheap transport of bulky raw materials from Lake Superior and vastly extending the space from which it was profitable to extract them (Temin 1964; Misa 1995; Priest 1996; Leitner 1998).

The locks were completed in 1855, just five years before the onset of the Civil War. Both sides used railroads extensively to move troops and matériel, and both sides used munitions and arms, the caliber, speed, and range of which had all been increased enormously by recent metallurgical improvements. The war tremendously expanded demand for iron and steel, both as armaments and to repair damage to the railroads. The war started, and was then conducted and finally won, as a struggle of manufacture and yeoman farming committed to union against the plantation agriculture and bound labor systems that were determined to secede. The migrant populations into the Ohio Valley and the northern midwest strongly supported union and, together with New England industry, profited from the huge demand for manufacturing and foodstuffs that the war created (Parker 1991).

Filling this demand greatly increased industrial capacity and accelerated depletion of the known deposits of iron ore. The locks at Sault Ste. Marie opened vast new territories that proved to contain the richest and largest iron deposits ever discovered. The physical and chemical composition of these ores and their shallow location made them particularly amenable to extraction with the new technological advances and adaptations made possible by recent technological innovations.

Just after the locks were completed—just before the Civil War—Bessemer in England and Kelly in the United States devised a much cheaper process for smelting large quantities of iron ore into steel, which was much stronger and less brittle than wrought iron. Lawsuits between Bessemer and Kelly delayed any production with the new technology in the United States until 1867, but once the contending parties pooled their patents, Bessemer plants increased rapidly in number, size, and quality of product.

The Bessemer process rapidly superseded the technologies that the Pennsylvania ironmasters had brought to the Ohio River Valley, but the proximity to high-quality coal deposits and river transport, combined with the increased scale and concentration brought about by coal-fired smelting, intensified the growing industrial agglomeration along that river. The new metallurgical technologies and the transport and agricultural innovations they made possible enhanced the region's manufacturing and trade, engendering new waves of migration to work in its increasingly diversified and growing economy.

The Civil War took a huge toll on traditional New England and middle Atlantic coast maritime shipping. This loss of external commercial opportunities sharpened the already intense east coast capitalist interest in the extractive and transport economies of the continental interior. The end of the Civil War and the huge industrial capacity and capital accumulation its conduct had spawned, the investment of this capital in the resumption of railroad building and in the rapid improvement and expansion of metal hulls and steam engines, the numerous technical improvements in the Bessemer technology, the frequent lengthening and deepening of the Sault Ste. Marie locks, and the continued discovery of large raw materials deposits (whose physical and chemical properties enabled and inspired further technical improvement) all converged to create the conditions for a major burst in the processes that the Bessemer plants made possible: the production of cheap steel of steadily improving quality.

The expanded consumption of iron ore quickened depletion of the Marquette and Menominee deposits in northern Michigan, but the Sault Ste. Marie locks and transit on Lake Superior opened access to the discovery of a series of larger deposits of higher quality; the Gogebic range between Wisconsin and Michigan was discovered in the 1860s, Minnesota's Vermillion range in the 1870s, and then the huge high-quality deposits of the Mesabi range in the 1890s.

Over the same period, trans-Appalachian exploration and settlement led to the discovery of extensive high-quality bituminous coal fields and limestone beds. Bituminous coal is more porous and less dense than anthracite. Its additional surface area makes it easier to combust and lets it burn at far higher temperatures than anthracite. Unlike anthracite, it can be cooked. This allows a far more regular and controllable burn. Bituminous coal enabled steelmakers to experiment with larger charges of iron and coal into larger smelters and furnaces operating at higher temperatures. This not only reduced fuel consumption, it also enabled steelmakers to produce stronger, less brittle rails.

This was particularly important, as railroad owners complained urgently about their rails cracking and breaking under the increased volume and weight of trains and cargos.

The increased scale of smelting created the demand for much larger boats, which the increased quality of steel provided material strong enough to build and operate. Maximum ore cargos increased from 1000 tons in the 1870s to 2000 tons with the first iron-hulled boats in 1882 to over 3000 tons when the first steel-hulled boats were built in 1886 (Priest 1996). Each increase in boat size impelled deepening and widening of the locks and encouraged the development of increasingly powerful loaders and movable conveyors. These machines reduced the time for loading cargos from the days that 1000 tons had taken in the 1870s to mere hours for over 3000 tons by the 1880s.

Technical improvements in smelting—together with the material properties of the less siliceous Michigan ores and the higher temperatures and greater control that bituminous coal allowed—reduced the furnace proportion of coal to iron ore, thus reducing the need for steel mills to be located near coal mines (Isard 1948). Cheap access to iron started to weigh more heavily in location decisions. Lake steamboat shipments cost less than a fifth of the per ton-mile rates for iron ore charged by the railroad. Steel mills soon spread from the upper reaches of the Ohio River—close to coal and water—to new locations along the south shores of Lake Michigan, from Detroit to Chicago, where boat cargos could be unloaded directly (Warren 1973; Priest 1996). Workshops for agricultural and mining tools and machines, yards for building steamboats and railroad cars, and, fifty years and many technical inventions and improvements later, the automobile industry's mass-producing plants followed closely (Parker 1991).

How Matter, Space, and Technology Shaped Mergers, Trusts, and the Multidivisional Corporation

Coal-driven smelting and the high heat requirements and capabilities of the Bessemer process led to increases in the scale and concentration of steel production. Scale increases in smelting made possible economies of scale in transport that exceeded the draft and length possible for riverboats. The size and depth of the Great Lakes, combined with the relatively flat topography around them, provided ideal conditions for an integrated system of rail and boat transport linking the iron mines northwest of the lakes with the coal mines to the southeast. Additional deposits were soon discovered in southwestern Illinois. The stronger, cheaper steel made it possible to build larger, stronger boats and trains, and the rapidly expanding national economy provided vibrant markets for steel and iron.

These linked cycles repeated on a far larger scale the interactions we have seen between space, matter, technology, finance, and politics as other economies have risen to global trade dominance. Each successive trade-dominant economy devised new technologies and combined them with the cheap and stable access to large volumes of raw materials that enabled greater and more rapid expansions in economies of scale.

As in the earlier cycles of material intensification and spatial expansion, realizing the new economies of scale required expanded forms of management and administration to coordinate the increasingly specialized divisions of labor and function. This, together with the greater infrastructural and technological costs of increased scale of production and transport, required expanded scales of investment, which could only be provided by new and larger financial institutions and instruments. Expanded financial and administrative systems, in turn, necessitated oversight, regulation, and coordination by more powerful, more competent state agencies able to exercise control over broader territories and more diversified economies. The increases in material transport and throughput combined with the need to coordinate multiple extractive, productive, and commercial processes across much greater spaces than had occurred before to engender new forms of property, administration, and finance and to combine them within unitary, highly complex organizations.

These evolved rapidly in late nineteenth-century America, progressing from judicial interpretations of the Fourteenth Amendment that endowed corporations with the same legal status as citizens to state laws allowing mergers and trusts between corporations. These legal innovations preceded major corporate and financial innovations at the end of the century that dealt simultaneously with the challenges and opportunities—vast material resources combined into complex systems of production, transport, and communication across vast spaces—presented to capital and with the huge scale of investment they required. The expanded investment requirements for technologies able to function on the enormous scale that American material and spatial scope made possible intersected with the more complex administrative challenges that expanded scale and scope engendered.

In America as in Holland and in Britain before it, the financial and administrative challenges shaped a solution appropriate to the markets and technologies of the time. America's iteration of this solution at the end of the nineteenth century was the emergence of the multidivisional corporation with the institutional support from and subsidization by state and national governments. At the same time, the national state gained in legitimacy and adjusted its legal and regulatory forms to moderate the threat that these corporations posed to competitive markets for labor and for goods. The organizational form of the multidivisional corporation, supported by state programs and policies to promote domestic development and to assure access to foreign raw materials and markets, facilitated further globalization.

To operate efficiently, the Bessemer process required coal of particular chemical composition and physical structure (with low proportions of carbon and moisture) as well as low-phosphorus iron ore, large quantities of limestone, and small volumes of high-quality manganese. The scale requirements, combined with the use of coal, facilitated the concentration of the iron industry (Temin 1964). The precisely specified physical and chemical composition required to maximize quality and efficiency, together with the extra costs of shutting down and adjusting the increasingly large and costly smelters and furnaces, made the uniformity of the raw materials extremely important. Each steel mill was dependent on predictable large deliveries on specific transport lines from specific large mines. These properties of the Bessemer process made possible and highly profitable the strategies of steel firm owners to increase capital concentration, spatial centralization, and vertical integration with the sources of raw material and transport.

Transforming vast quantities of cheap steel into trains, rails, steamboats, docks, cranes, and derricks enabled cheap and reliable delivery of heavy cargos of iron, coal, and limestone to the increasingly centralized and rapidly growing steel plants. The same transport systems also enabled the equally crucial distribution of finished steel and steel products to the industrial centers emerging across the national territory. They carried the swelling agricultural production produced with cheap steel-based machines and tools to the midwest's rapidly growing urban workforce. As the machine and tool industries grew, they— along with the railroads—put increasing pressure on the steel mills to standardize and unify their products. Standardization of output provided a further advantage to concentration, centralization, and scale.

Improvements in the Bessemer process and in rail-making process; the construction of rail lines, steamboats, bridges and the first skyscrapers; and the flourishing market for agricultural implements all supported further economies of scale in steel mills. The improved rails and stronger trains supported the movement of greater volumes over greater distances—from mine to centralized smelter and then to markets across the entire national territory—and railroad companies profited from the additional ton-miles the steel industry required.

The discovery of enormous deposits of high-quality, soft iron ore near the surface of the Mesabi range enabled further technological developments just as demand for rails was starting to slow and demand for higher quality slabs and plates suitable for construction of buildings, ships, vehicles, and other steel products was growing. Open-hearth processing produced higher-grade steel than Bessemer smelters could manage, but at first it was much costlier because of the smaller volume of its charge. The combination of Mesabi ores and bituminous coal and coke allowed for larger charges and a steady reduction in operating costs. These materially enabled technical improvements expanded the supply of cheaper steel of higher quality. Abundant supplies of stronger, cheaper steel encouraged the development of new technologies and products, diversifying and expanding demand for steel and generating multiple other industries downstream. The physical properties of Mesabi ore facilitated mechanization and economies of scale in extraction via less labor-intensive surface mining and transport as well (Priest 1996: 81).

Profits for the vast capital sunk in the rapidly growing and increasingly complex steel industry depended on reliable links between the multiple coal mines south of the Great Lakes, the multiple iron mines north of the Great Lakes, the vast beds of limestone, and the multiple rail and steamboat lines. Through the 1880s and 1890s, J. P. Morgan, Andrew Carnegie, and a few other powerful financiers and industrialists started to purchase and consolidate different parts of this complex, usually combining control of a mine with control of its transport system. During this same period, John D. Rockefeller used the skills and capital he accumulated in negotiating discounts and rebates from the railroads that shipped his oil to buy up the growing and ever busier fleet of Great Lake ore and coal carriers.

Morgan invested extensively in railroads as well, particularly eager to avoid competition that might result in wasteful duplication of rail lines between the mines and the markets. He bought a controlling interest in various mines and railroads to avoid such outcomes. Carnegie formed a company that merged several iron mines, coal mines, and diverse railroads and steamboat lines with his own steel mills. Each mill was located on a major river that allowed barge transport for iron ore and was fed by tributary rivers through the vast, high-quality Connellsville bituminous coal fields on the western slope of the Appalachians. This enabled the mills to combine cheap transport of Great Lakes iron with cheap transport of bituminous coal, whose light and powdery form made it easy to combust but difficult to transport without breakage and collapse. Carnegie then merged his company with Rockefeller's company, combining the nation's largest steel company with vertically integrated control over all of the iron, coal, coke, railroad, and lake transport that it needed to operate. This merger tremendously lowered Carnegie's material and transport costs—and frightened his competitors into a wave of similar, though necessarily smaller, mergers.

In 1901, Morgan and Carnegie agreed to end their "destructive competition." Their agreement created U.S. Steel—the largest corporation in the world up to that time—by merging their holdings in transport, raw materials, and processing with a number of smaller transport, mining, and smelting firms. The spatial dispersal of the natural sources of this corporation's varied raw materials, together with its huge capital stock, made U.S. Steel the quintessential example of that great American innovation in economic organization—the multidivisional corporation.

Arrighi (1994) follows Chandler and most other economic analysts in identifying the reduction of transaction costs as the multidivisional corporation's major purpose and economic effect. Reduced transaction costs do clearly result from any form of vertical integration. Fully appreciating the economic effects of the multidivisional corporation on both the U.S. and global economies, however, requires us to take into account as well its solutions to the challenges of financing and coordinating accelerated material intensification across rapidly expanded space.

Moving vast volumes of cargo across vast spaces from multiple origins to centralized destinations without investing in costly parallel rail lines presented huge and complex problems. Frequent passages on single lines of track in opposing directions required close coordination of multiple trains on multiply overlapping trajectories with huge downside costs in life, equipment, and cargo for errors or oversights. Chandler (1977) describes the origins of the multidivisional corporation as an evolving, then generalized, solution to this transport problem. The multidivisional corporation manages numerous separate operational units widely distributed in space and autonomous in many local decisions but subject to policy control by a central headquarters and bound to technically precise procedures of accountability and coordination. Chandler first traces the railroad industry's development and perfection of this managerial form and then shows how it spread to other industries.

His account of how the telegraph developed in interaction with the railroad's complex systems of coordination and control for its spatially and operationally separate units vividly illustrates the intimate entwining of technical and organizational innovation as both forces respond to the complex material and spatial challenges of increasing the scale and scope of production. Chandler does not, however, recognize that the multidivisional corporation developed in response to the historically and territorially specific opportunities and challenges confronting nineteenth-century America—vast space containing huge and varied materials incorporated into the world economy as the technologies appropriate to those spaces and materials were being developed. Nor does he recognize the multidivisional corporations' utility in combining multiple separate capitals but allowing sufficient autonomy between the component divisions to allay somewhat the distrust and opposition between individual competing capitalists.

J. P. Morgan adapted and extended the multidivisional corporation to deal with the immense capital barriers to vertically integrating huge mines, huge smelters, and vast transport systems on the one hand and the personal antipathy and distrust between himself, Carnegie, and the other major shareholders and owners on the other hand. He combined their diverse holdings of mining and transport interests into a single merged company, United States Steel, in which some of the individual corporate owners maintained some autonomous control of the separate component parts.

The nation's investors were tremendously excited about the prospects of the expanding iron, steel, and transport industries within the booming U.S. economy. When U.S. Steel shares were publicly offered, so many investors sold other stocks to buy shares in steel that the entire New York Stock Exchange crashed as share prices of other firms fell dramatically. Morgan found an effective way to collectivize capital in a single huge project that would solve the problems of coordinating huge material flows across vast spaces. The financial motives and results of his negotiations were as consequential for the national economy as the organizational revolution they achieved, but the broad spreading of risk through the coordination of multiple highly concentrated capitals is seldom acknowledged in analyses of the multidivisional corporation.

Merging the interests of many separate capitals provided huge advantages of scale to the country's most dynamic industries. At the same time, it threatened to enable the merged transport and raw materials industries to restrain trade for competitors in ways that would ultimately damage the national economy. Agricultural interests had already organized against the abuse of transport monopolies in shipping grains and meat to market. Some of the most successful and politically radical labor unions were organized in those sectors of heavy extraction and industry—coal, iron, steel, and railroads—most affected by the merger movements that had led to the multidivisional corporation. Labor organizers discovered that the growing articulation and interdependence of specialized production units with the multidivisional corporation enabled them to coordinate strikes whose effects on both the corporate and the national economy extended well beyond the particular unit in question.

The national state therefore had to invent new forms of regulation as powerful as the new corporate forms but sufficiently subtle and flexible to impede restraint of trade without unduly slowing capital accumulation. It also had to find new ways to mediate between the growing power of capital and the growing militance of labor and agriculture. Through the entire twentieth century, capital, labor, and the state struggled through a series of lengthy negotiations within the framework of extensive litigation in a process of ongoing learning and compromise that continues today.

From Continental to Global Expansion of Economic and Political Power

The foundation of U.S. Steel coincided with the slowing of railroad construction, the closer economic and political articulation of the still unevenly developed regions of the nation that the railroads made possible, the spread of the open-hearth furnace and of the multiple new structural applications of the better steel it produced, and growing political interest in extending U.S. economic and political power beyond its continental territory to other territories where American capitalists wanted favorable trade. These varied conditions and accomplishments concatenated to promote the use of steel in huge fleets of steel-built, coal-powered battleships and gunboats to establish a significant U.S. naval presence on world oceans during the first half of the twentieth century and increasing proportions of U.S. direct investment in foreign economies during its second half.

The basic open-hearth furnace took any iron ore with a phosphorus content up to 1 percent, making "available immense iron ore deposits which could not be utilized otherwise of their phosphorous content . . . too high to permit their use in the acid Bessemer or acid open hearth processes and too low for use in the basic Bessemer process" (U.S. Steel technical manual, cited in Priest 1996: 75). Carnegie built the basic open-hearth furnace into his Homestead mill in 1888. Experimentation with different types of coal and iron and with the temperature and duration of the blast soon enabled larger charges in larger furnaces, greater speed of the burn, and more precisely controlled shape, thickness, and hardness. These advances lowered open-hearth costs, improved steel quality and versatility, and expanded steel's uses just as the petroleum-powered internal combustion engine opened new possibilities for technological innovations in production and in transport. By 1908, open-hearth production surpassed Bessemer production; by 1912 it had doubled it.

The combination of open-hearth steel and the internal combustion engine opened the way for an extensive and flexible system of highways and vehicles. These in turn extended the production and exchange of bulky goods into the great spaces between the more rigid rail and river lines. Highways and the smaller, lighter vehicles that ran on them were far less costly per mile traveled than rails and canals and the larger, heavier trains and boats that they carried. These cheaper, lighter modes of transport could thus operate within and between the lines that needed economies of scale to repay their much greater construction and operating costs. Highways thus complemented the railway system's economic expansion across space with a deepening, or thickening, in multiple connected places. The tremendous economic contribution of small, lightweight, and flexible means of powered transport stimulated further experimentation and development of efficient lightweight engines that powered faster, stronger, heavier vehicles, boats, and eventually airplanes (Hugill 1994). Rapidly improving diesel engines and stronger open-hearth steel increased the flows of materials into and within the United States to vastly greater levels than any other national economy had ever achieved, while at the same time enabling extension of its growing economic and political power beyond its territorial boundaries.

The expansionist impulses that drove the incorporation of the American midwest and far west into the rapidly globalizing world economy had also made the United States the most productive economy and one of the world's wealthiest and most powerful nations. There were now equally strong impulses to mobilize this wealth and power for continued expansion beyond the North

American continental limits, though there was also considerable opposition to the projected cost of the proposed projects. These debates continued in changing form from the turn of the century until World War II. They may have moderated, but did not stop, a series of collaborations between capital and the state to extend U.S. economic, financial, and military power abroad.

Among the many new uses of improved open-hearth steel that replaced the dwindling demand for Bessemer rails was the construction of a large fleet of huge battleships, cruisers, and destroyers that the U.S. government built during the first three decades of the new century. Stronger, lighter steel made possible the larger, stronger plates for hulls and the large-caliber, long-range guns that equipped these ships. The so-called Great White Fleet embodied U.S. policy and strategy to extend its economic and political power beyond its continental territory. Military demand for huge quantities of very high-quality steel encouraged further technological improvements and paid for their implementation.

The national state's executive branch, influenced by naval strategist Alfred Thayer Mahan, patrician imperialists like Henry Cabot Lodge and Theodore Roosevelt, and patriotic intellectuals like Frederick Jackson Turner, proposed aggressive expansion into the Pacific and Caribbean (Zimmerman 2002). In 1893, Turner first formulated his thesis that America had formed its essence by pushing back the western frontier. These ideas strongly influenced Theodore Roosevelt and supported his strong commitment to incorporate Cuba into U.S. commercial and political control. The executive branch of government was more inclined than either business or the Congress (whose members were more sensitive to business interests) to undertake colonial conquest (Zakaria 1998), but business interests and Congress were both quite happy to have naval protection of their growing international investments and commerce, so they supported naval expansion.

Annexation of Mexican territory and then the California gold rush in the 1840s had long since captured east coast commercial interest and sparked a series of schemes to expand transport and commerce in the Pacific. Cornelius Vanderbilt and a group of New York associates had by 1849 procured a Nicaraguan concession for a ship canal, and soon after that they connected their Atlantic and Pacific steamship lines by roads and lake boats. Hawaii had been the base of the U.S. whaling fleet in the Pacific since the 1820s. By the 1850s, commercial routes linked the Puget Sound, the Columbia River, San Francisco, and San Diego to ports in Mexico, Peru, Chile, Tahiti, Australia, and New Zealand (Meinig 1993). Shows of naval power opened Chinese ports to American ships in the 1840s; Japanese ports followed in the 1850s. Again, and even more dramatically than in shorter ocean crossings, improved transport technologies—from clipper ships to steam boats—moved the Pacific trade from exclusively high-value/low volume trade (whale oil, furs, tea, and silk) to lower-value/higher-volume trade (wheat, flour, rice, lumber, and hides). By the turn of the century, diesel-powered internal combustion engines drove the far larger open-hearth steel ships in the hugely expanded U.S. fleets, both merchant marine and naval, to the Pacific.

Despite the technological improvements and size increase of its naval ships, America's self-sufficiency in the bulkiest raw materials restrained investment in the carrying capacity of the U.S. merchant marine; as a result, the average size of individual U.S. cargo ships and the total size of the U.S. merchant fleet remained smaller than Britain's until World War I. The greater flexibility and shorter overland distances enabled American national rail lines, lake boats, and river barges, increasingly complemented by trucks, to compete with international ocean transport's far greater economies of scale.

The German merchant fleet, in contrast, grew rapidly from 1900 to 1914 and, with the support of major banks and the state, formed the centerpiece for an assault on British colonial and commercial dominance. World War I and the subsequent distribution of the German fleet among the victorious nations ended this threat, but between the world wars British carrying capacity stagnated while the American, Japanese, Norwegian, and Italian fleets all increased two- to fourfold (Cafruny 1987: 60–65).

The massive industrial production for World War I occasioned serious shortages of critical materials such as manganese, tin, and rubber, leading the U.S. state and industrial capital to formulate access strategies for foreign sources (Staley 1937; Eckes 1979; Bunker and O'Hearn 1992; J. Marshall 1995). Still, though, the bulkiest raw materials—coal, iron, and oil—were all accessible overland, if not in the United States, then in Mexico and Canada. The U.S. merchant fleet grew, but less than gross national product did, and the sizes of the largest ships were still set by the drafts possible on the Great Lakes.

World War II vastly expanded ship production, which resulted in huge excess carrying capacity at the end of the war. These boats were still scaled to the Great Lakes traffic, though. The huge overhang in supply impeded development of larger boats, even as the United States started importing iron and oil from Latin America. As the United States depleted some of its raw materials, digging through the best iron on the Mesabi and importing increasing amounts of oil from Venezuela and then from the Middle East, and continued to export vast amounts of food in the 1950s and 1960s, U.S. merchant ships did not expand past about 60,000 deadweight ton (dwt) cargos that had served so well on the much shorter Great Lakes routes.

The Germans and the Japanese did not enjoy overland access to raw materials—in particular to oil—that the United States had enjoyed, so they put much more effort into developing their maritime capacity. German expansion before World War I and Japanese expansion before World War II show interesting parallels with British strategies in the seventeenth century (see also Hugill 1994). In all three cases, the state undertook massive support for the development of transport systems that aimed to secure access to raw materials through challenging competitors' existing market and transport dominance. In all three cases, the international tensions that resulted led to major wars. The United States was exceptional in that massive state support for constructing transport systems to secure access to raw materials could be sunk in unchallenged national territory rather than in international waters. Its production system was efficient enough and its raw material supplies were self-sufficient enough that it could choose to leave much of the international carriage of raw materials to the merchant fleets of less wealthy nations.

The raw materials like rubber, tin, and manganese that the United States needed to import all had too high a value-to-volume ratio to justify major research and investment in transport economies of scale. Business and diplomatic efforts such as the Grand Area strategies of the Council on Foreign Relations (CFR), Truman's Point Four program, the conditions imposed on receipt of Marshall Plan credits (Bunker and O'Hearn 1992), and the Trilateral Commission (Girvan 1980) tended instead to focus on internal transport in their nations of origin and on security of access and passage for U.S. companies. These programs all extended strategic principles the CFR had first propounded in the 1930s: U.S. access to foreign raw materials could be most cheaply and securely provided if the economies of the source nations developed sufficiently to install and manage the necessary extractive and transport infrastructure (Staley 1937; Bidwell 1958; Bunker and O'Hearn 1992). Such economic development would also expand these nations' demands for U.S. industrial and agricultural products, which could be cheaply exported, serving as back-haul freight on the transport systems built for their raw materials exports.

These principles legitimated U.S. state acceptance of and implicit support for the Import Substitution Industrialization (ISI) policies and programs that many Latin American and Asian national states adopted during the 1950s and 1960s, even though earlier hegemonic regimes would have perceived the tariff barriers they provided for their "infant industries" as a threat to trade (see Girvan 1980; Bunker and O'Hearn 1992; O'Hearn 2002). At the same time, U.S. capital extended the logic of the multidivisional corporation to overseas branches. This transformed the multidivisional corporation into the transnational corporation (TNC), a form that used cheaper foreign labor and local inputs to produce finished goods, both for local consumption and for shipment back to the U.S. market. Local U.S.-controlled subsidiaries of TNCs often benefited from the state subsidies and protective tariffs that ISI-promoting foreign states had put in place.

For all of these reasons, after the transition from wood to iron as the world's most voluminously used industrial input, shipping was not as critical to competitive production, nor was it as effective a generative sector, in the United States as it had been in other hegemonic economies. Canals, steam-driven riverboats, railroads, and then highways and trucks provided the transport that opened the vast and resource-rich internal American space. From the 1830s onward, the U.S. economy became progressively far more articulated internally than it was oriented to external commerce. Agriculture and extraction continued to provide a very large share of what it did export, even after it became the world's most productive and wealthiest industrial economy. Unlike industry, agriculture and extraction are horizontally dispersed, so their export depends first on overland transport. The overland transport systems in the United States and the ways they solved the nation's particular iteration of the contradictions between economies of scale and the cost of space were far more generative for the national economy than international shipping and shipbuilding.

The extraordinary success of this internal or continental self-reliance encouraged U.S. military and diplomatic planners after World War II to promote its allies' industrial development, sacrificing U.S. trade dominance to geopolitical tactics aimed at building a strong capitalist bulwark against the spread of communism south from the Soviet Union into Europe and Asia (Bunker and Ciccantell, forthcoming). The sustained success of labor's insistence that it share in the country's rapidly growing wealth had by the 1950s spawned an extraordinarily high per capita consumption of raw materials. These high consumption levels sustained the continued rapid growth of the U.S. economy, but they also made the United States increasingly dependent on foreign raw materials, particularly oil but also aluminum and other specialty metals.

U.S. foreign policy used this vibrant consumer market and the extraordinarily vast capital accumulation that supported it to foster manufacturing throughout the portion of the world aligned with it in the Cold War. U.S. corporations exported to the periphery technologies and economies of scale that had been surpassed in the core. They thus extended the surplus profits they gained at the moment of their innovation by installing them in areas where they could purchase cheaper labor and raw materials. Global industrial production and natural resource extraction expanded rapidly, but U.S. consumption of both natural and social products expanded far faster than the global average. Paradoxically, while U.S. industrial production grew less rapidly than global industry, U.S. agricultural production and exports—aided by high returns to labor and capital and national state policies and subsidies—continued to grow, in both absolute and relative terms. Continued state support for transport contributed significantly to this outcome by making it easy and cheap for the United States to integrate its land-extensive agriculture with international markets.

Labor and capital were both served by the cheap transport of cheap raw materials, and the national state acquiesced to their combined pressure by subsidizing highway construction, keeping gas and oil taxes low for both producers and consumers, providing depletion subsidies at home and political and tactical support for oil companies abroad, and applying antitrust legislation sparingly. The U.S. highway transport industry—from automobile manufacturing to trucking—flourished, as did the U.S. oil industry, even as dependence on foreign oil increased.

Other raw materials–based industries declined as their domestic resource bases diminished. Even the mighty steel industry was affected. Bethlehem Steel had established its mills at Sparrows Point on the Chesapeake Bay to cheapen transport from its own South American iron mines. In the early 1950s, U.S. Steel invested hugely in new technologies to mine the very hard taconite ores that remained in the Mesabi after miners had taken the more easily mined and processed hematite ores there. However, within a decade participants in the national economies that the United States was subsidizing had discovered new mines and invented new maritime technologies that left the Mesabi ores uncompetitive, the vast U.S. Steel mining and transport system redundant, and its lake-locked steel mills poorly located. Imports of steel increased as well, particularly on the west coast. The relative decline of heavy industry in the United States tremendously weakened the labor unions, while capital prospered from its increased global mobility. The U.S. gross domestic product grew rapidly through the 1980s and 1990s, but income and wealth became far more concentrated.

British-American Complementarity in Comparative Perspective

The complex relations between the British industrial and political core and the American raw materials–supplying periphery reiterated, in greatly expanded and prolonged form, the intensification of material flows across broader spaces and the materio-spatial and technological interdependence between core and periphery that had marked earlier transitions between trade-dominant nations and the systemic cycles of accumulation they led.

Until the transition to iron and coal technologies, the materio-spatial interdependence of established hegemon and rising challenger consistently ended much more rapidly than British-American interdependence did. In contrast, Britain continued to devise new technologies and the social and financial organization and institutions needed to implement and operate them in sustained interaction with America even *after* the United States achieved independence and rapid industrial growth. The U.S. economy, in turn, adapted these technologies to expand further its material sources by opening and cheapening access to broader territories in its internal space. It then improved these technologies to enhance labor productivity and to reduce the costs of extraction and transport. This enabled the forestry, agricultural, and mining sectors to supply raw materials in increasing volume and diversity for rapidly expanding domestic and European industry and to supply food for the rapidly growing urban work force of both continents. British capital collaborated with the British state to finance the canals and railroads that were opening the American interior to larger shipments of lower value-to-volume products.

America's raw materials exports to Europe continued to grow while its manufacturing—and its exports from manufacturing—grew even faster. Each of the linked growth cycles generated self-reinforcing synergies and feedback loops connecting technological innovation, the incorporation of new spaces, the discovery of new types and volumes of raw material, and the development of the political and financial institutions required to govern and finance the expanded forces of extraction and production. Thus, innovating core and industrializing raw materials supplier each complemented and supported the other's sustained development of progressive economies of scale and economic diversification, even as the U.S. rates of production and growth consistently surpassed Britain's after the 1880s.

The flow of raw materials to Britain and of technology and finance to America shaped and was shaped by industry's transition from primary material dependence on wood to primary material dependence on iron and coal. The two nations played different, but central, roles in that transition. Wood technologies were limited to animal traction or wind for motive power. This constrained commerce in bulky materials to goods produced near the shores of navigable rivers, lakes, and oceans. Afterward, iron and coal technologies provided more efficient and more versatile motive power that opened continental interiors to increasingly high-volume/low-value commerce. Railroads and steamboats freed the trade and transport of cheap, bulky goods from its reliance on downriver flows to coastal harbors. Metal hulls, engines, shafts, and propellers enabled progressively larger, cheaper, and faster maritime transit.

America turned coal and steel transport technologies inward, building railroads and steamboats for commerce in its interior spaces. Lacking similarly large and easily transitable internal spaces, Britain turned outward to control international shipping, finance, and insurance from its pivotal position between the Atlantic crossing to America and the Indian Ocean connection to Asia. America used its own raw materials to mechanize agriculture, forestry, and mining, while Britain imported raw materials and transformed them into capital goods and commodities for its own market and for international markets.

Britain's strength in transmaritime commerce articulated well with America's growing industrial power and internal transport capacity. Their conjoint economic expansion tremendously accelerated and deepened globalization. It also strengthened the ability of both nations to control global trade. By the middle of the nineteenth century, expanded global trade in increasingly high-volume/low-value raw materials created the conditions for the British economy to create a "second industrial revolution" in which it made "machines that made machines." The near-monopoly of British-produced capital goods in international markets made that country the "workshop of the world."

The continued British-American reciprocity of raw materials supply and technological innovation—already hugely intensified as iron and coal replaced wood as the most voluminously consumed industrial raw materials—deepened and extended even further with the development of the petroleum-driven internal combustion engine. Petroleum delivers more energy per pound than coal. The machinery needed to convert this energy to mechanical power is lighter and smaller, and therefore uses less cargo space, than the machinery and water required to convert the energy in coal to steam, contain the steam under pressure, and then convert the force of steam-pressure into mechanical energy. These material differences between oil and coal and their respective technologies made possible the far greater flexibility and agility of smaller vehicles on roads rather than on rails. The lower capital costs of each transport unit allowed for mass ownership, and therefore mass production, of gas-powered vehicles. These material effects converged with new chemical technologies that made fertilizers cheaper to fabricate, transport, and apply. Cheap fertilizers and flexible transport opened new areas of American space to commercial agriculture and extraction. The growing military and economic importance of petroleum impelled Britain to extend its imperial controls into areas of the Middle East it had previously seen as useless desert.

It also tied U.S. agricultural and industrial development even more tightly, both internally to each other and externally to Britain's imperial economy. The internal combustion engine and the contemporaneous development of electrical generation, transmission, and conversion to mechanical power vastly expanded both the technologies of production and the range of commodities that could be produced. Indeed, both forms of energy made it possible to produce machines—from automobiles to vacuum cleaners—as end products as well as units of production. The utility of the new technologies often rested on small, mobile units delivering powerful energy. Building small, light, mobile engines of great power required a much greater variety of material components such as rubber, tin, and copper. Many of these components were most easily available in Africa, Latin America, and Asia, all places distant from Britain and the United States.

The new technologies enabled bulk transport of raw materials from all of these places, while growing dependence on these raw materials encouraged Europeans and Americans to use the new transport technologies for military and colonial conquest and administration. From the middle of the nineteenth century onward, steamboats, railroads, later diesel-powered ships, and finally automobiles, trucks, and airplanes facilitated core control of diverse continental interiors that sent essential raw materials back to Britain and to America. European colonial administrations, especially Britain's, adapted American riverboat technology, its improvements in railroads, and its advances in internal combustion engines to extend core control and commercial access from coastal cities to interior territories (Headrick 1981). The new transport technologies simultaneously diffused and strengthened the military and administrative systems set up to secure access to a broad range of cheap, bulky raw materials that were increasingly critical as industrial inputs and as cheap food for urban labor.

Steamboats and railroads did not just transport tin and rubber from Malaya and Indonesia; timber and ore from the Philippines; rubber from the Belgian Congo; cotton from Egypt, India, and East Africa; copper, nitrates, tin, and oil from nominally independent nations in Latin America; and grains from around the world. They also enabled a high-volume trade in coffee, tea, and sugar—all of which had been luxury goods before iron and coal cheapened and accelerated transport and turned them into staples of mass consumption (Mintz 1986). The invention of electrically refrigerated rail cars and ships' holds in the late nineteenth century eventually connected the Great Plains of the United States (and soon afterwards the vast Argentinian pampas, New Zealand pastures, the Australian outback, and North Atlantic and Southern Pacific fisheries) to markets in the east coast and later in Europe. This made fresh meat and fresh fish cheap enough for urban working-class consumption, further reducing the costs and increasing the profits of production and trade.

Technically driven expansion of mass consumption markets accelerated during the twentieth century. The parallel development of the internal combustion engine and the electric motor—steadily improved and cheapened by new and strong lightweight metals and more precise machine-tooling techniques—opened and expanded whole new markets that vastly expanded the ranges and volumes of raw materials incorporated first into the machinery of mass production and then into the increasing range of commodities cheap enough for mass consumption. The rising affluence of core workers and the growth of the middle class in many peripheral nations supported the increasing globalization of these markets and the increasingly concentrated profits that they provided to capital.

The rapid and sustained growth in technical and economic scale that mineral technologies made possible during the long series of British-American exchanges generated new global forms of finance, politics, markets, and class relations. The core nations' increased dependence on the growing volumes and increased range of raw materials across broader and broader spaces demanded both technical efficiencies of extraction and transport and secure international regimes of finance and administration. Thus, iron and coal, later supplemented by oil, enabled new economies of scale that required greater volumes and varieties of raw materials from a growing number of increasingly distant sources. Britain and then the United States constructed new institutions and instruments of international finance, managed by increasingly convergent national state structures, to facilitate, secure, and coordinate these increasingly complex, voluminous, and diversified flows of matter across greater and greater distances and a more diverse set of sovereignties. Once again, the interacting opportunities and imperatives of accelerating technological scale quickened, extended, and deepened globalizing processes. The reiterated, steadily expanding solutions to the contradiction between economies of scale and the diseconomies of space drove progressive globalization: first, of raw materials procurement and of the transport systems it required; then of the financial and political systems needed to cheapen, secure, and regularize raw materials markets and the transport systems they required; and finally of the re-externalized production systems needed to satisfy the booming market for consumer goods that sustained the globalizing economies of scale.

The uniquely prolonged reciprocal catalysis of economic growth between the United States and Britain emerged from the very particular material and spatial properties of each as changing technology and extensive infrastructural construction mediated each society's interaction with their natural attributes. Britain opened the Suez Canal in 1869, the same year that the United States completed the transcontinental railroad. The old dream of a westward route to China had finally been realized by a raw materials periphery rather than by the dominant European nation. There were now two efficient routes to East Asia.

The realization of the circum-global trade these routes made possible rested on the continually expanding and diversifying exchange of technologies and raw materials between Britain and the United States and on the articulation of American continental steamboat and railroad lines with Britain's transoceanic shipping. The two nations collaborated technologically and administratively in maintaining spatially and materially diverse supply chains. The Suez Canal opened cheap eastward transport to East Asia, and the railroad across the American continent was a westward means of achieving the same thing. Expanding rail and steam systems in Latin America and Asia provided both nations with cheap and stable access to raw materials and the means to rapidly deploy administrative personnel and military force against any challenge to their growing power.

The respective locations of the United States and Britain in relation to the Atlantic, Pacific, and Indian Oceans put the two nations in coordinated control of the two routes and enhanced their economic and imperial interdependence. The railroad accelerated the fuller incorporation of America's western frontiers. The new agricultural and extractive opportunities attracted migrants who rapidly turned western territories into states. Having incorporated its entire internal territory, the United States waged war against Spain in Cuba, moved to secure Hawaii and the Philippines, and fomented Panamanian secession from Colombia to secure solitary control over the Panama Canal it built to connect its Atlantic and Pacific zones of commercial and naval supremacy. It used these territorial and maritime expansions for military bases, coaling stations, additional sources of raw materials, and protection of its expanding merchant fleet against the growing presence of German and Japanese navies in the Pacific.

Except in Hawaii, Puerto Rico, a few remote Pacific islands, and for a limited time the Philippines, the United States did not attempt to imitate British imperial strategies of direct territorial control. Even in Panama, it directly administered only the land required to assure maximum security for the strategic infrastructure of the Panama Canal, while saving money and good will by leaving the administration and defense of the national territory to a national state whose putative independence it had helped secure. In this, it secured a huge advantage over its early twentieth-century competitors for trade dominance, Japan and Germany, who were attempting to apply outmoded nineteenth-century British colonial strategies to the far more sophisticated, nationalistic, materially more intense, and spatially more closely connected twentieth-century global economies. Lacking the immense domestic resources of the United States, Japan and Germany had few options for sustaining economic growth and increasing their political power. Imperial expansion became a costly but apparently necessary strategy. However, Japanese and German strategies provoked two world wars, both of which they lost.

The sustained reciprocal dynamic of the economic power of both nations reached the pinnacle of its globalizing effects early in the twentieth century, when they achieved the full incorporation of the Atlantic, Pacific, and Indian Ocean economies and of the vast continental interiors behind their shores. British-American interdependence by then had evolved into a global triangular trade. Britain imported American manufactures that it paid for with funds collected from the raw materials—first and foremost rubber, but also tin, silk, rice, coffee, and tea—that its African and Asian colonies provided to the American economy (J. Marshall 1995). The former extractive economy now imported raw materials for its booming industry from its erstwhile imperial patron's remaining colonies. The United States devised new ways to dominate those colonies after it had successfully advocated their formal independence, using the Marshall Plan and other forms of persuasion to gain rights to invest directly in (and extract raw materials from) them in the aftermath of World War II (Bunker and O'Hearn 1992). More than three centuries of evolving British-American exchanges of raw materials and technology, across the Atlantic and eventually around the world, had by then alternately extended integrated commerce across more distant oceans and then incorporated growing portions of the continental interiors beyond foreign shores.

CHAPTER SIX

Raw Materials and Transport in the Economic Ascendancy of Japan

Japan's rapid rise to trade dominance in the late twentieth century globalized raw materials markets and the bulk transport of extremely low value-to-volume cargos more dramatically than any previous national economic ascent had done. Its rise, though, was far more conditioned by geopolitical forces beyond its control than any previous rise had been, so whatever dominance it achieved was extremely precarious. Japan's challenge to world-system hegemony appears to have ended in economic crisis, but the globalizing effects of its economic strategies endure, as do the inequalities they exacerbated.

These strategies, and their consequences for Japan and for the peripheral countries that supply its raw materials, raise disturbing questions about the reiterated solutions to the contradictions we have described between natural and social production and between economies of scale and diseconomies of space. First, can the self-reinforcing and sequentially cumulative cycles of economies of scale that create diseconomies of space solvable only by even greater economies of scale continue as the searches for new raw materials sources reach global limits? Second, what will be the long-term consequences of the accelerating social and ecological disruption and inequality that are caused by globalizing raw materials extraction and transport? We will address these questions after we examine how shipbuilding, shipping, and their use to procure competitively cheap access to iron and coal catalyzed historically unprecedented levels of collaboration between the state, heavy industrial firms, their sectoral associations, and financial institutions. We will see how these collaborations crucially supported Japan's rise to trade dominance. In this sense, the procurement and transport of the raw materials most voluminously used in contemporary industry constituted for Japan, as they had for Holland, Britain, and the United States, the generative sectors that drove national ascent to trade dominance while pushing the world economy further along the path of globalization.

Transport as Japan's Solution to Geopolitically Enhanced Costs of Space

We have stressed the parallels in the ways that privileged access to the raw materials most used in the construction of bulk transport vehicles structured both the internal and the world organization of rising economies, but it is important to remember as well that the material intensifications and spatial expansions that drive each cycle of trade dominance accumulate. They generate a rapidly accelerating world capitalist system that progressively incorporates a widening area into ever closer and more complex integrations of production and exchange. Japan's rise is the most recent of a series of reiterated cycles, each of which has extended the scope and depth of the world economy through massive increases in the scale of capital, the adoption of more powerful technologies, the industrialization of new populations, and the creation of new financial and political organizations with powers proportionate to the new scales and the broader scope of the world economy.

Significant reorganization of the world-system's hierarchy occurs infrequently and in ways that reflect the political and economic organization of previous hegemons. As technologies, financial institutions, and markets change over time, different ascendant economies working out of different materiospatial contexts confront very different opportunities and constraints. At the same time, in creating and implementing both sectoral and national strategies to enhance its competitive position, each ascendant economy creates new technologies and financial institutions and contributes to the growth of production, capital, and labor in the world as a whole. These effects change the obstacles to and opportunities for their strategies to which they responded in the first instance. Success in ascent and endurance as a trade-dominant nation depends on the ability to respond to changes that one's own actions are catalyzing.

Japan, the most recent of the rising economies we consider here, devised strategies appropriate to a larger, spatially more dispersed world economy than any of the earlier ascendant economies. At the same time, Japan's particular situation at the beginning of her ascent—that is, extreme resource dependency combined with great distance from available sources under the aegis of the apparently secure U.S. hegemony—forced its decision-makers to extend dramatically the space of the world economy they confronted as they began their rise and then to adjust to the changes their strategies engendered within the world-system.

This created some important departures from the trajectories of earlier trade-dominant nations that had started with advantageous proximity to the raw materials they used in greatest volume. The Dutch transformed their advantageous access to wood into a highly competitive carrying efficiency, out of which Amsterdam became an entrepôt of warehousing, trade, and letters of exchange. British access to coal and then American access to coal and iron enabled both nations to achieve tremendous economies of scale in production, which were then expanded through global control of currency and trade. The British use of imperial control was supplanted by American systems of direct investment abroad on a massive scale. Japan, in contrast, promoted increases in the scale of ports, ships, and integrated transport systems to reduce the huge costs of distance to politically acceptable raw materials sources in the context of the Cold War and U.S. support for Japanese industrial development after World War II. The cost of this transport-based strategy was so high that it could only succeed by devolving much of the cost onto other economies. Japan's success in doing so, though, expanded global transport systems at a rate the world had not seen since the rapid increase in transport scale that British steamships had led at the turn of the twentieth century. In order to compete at all, Japanese industry had to overcome and then surpass the barriers to entry earlier levels of scale imposed. In both the steel industry and shipbuilding, Japan fairly quickly took the lead, not simply in adapting to the optimal scales in each industry but in pioneering the innovations that increased those optimal scales (O'Brien 1992; Chida and Davies 1990).

These increases in scale were phenomenal, and they required vastly expanded capital investments. In steelmaking, the minimum efficient scale of an integrated steel plant rose from about 2.5 million tons of capacity in 1950 to about 7 million tons in 1970. The optimal capacity of a blast furnace increased tenfold in this period. In 1950, the largest tankers ranged from 10,000 to 20,000 deadweight ton (dwt). By 1968, the largest tankers were over 300,000 dwt. As recently as 1971, 65 percent of iron ore was transported in ships of less than 60,000 dwt, and only 11 percent in ships of 100,000 or more dwt. By 1983, only 14 percent of iron ore was transported in the smaller ships, and ships of over 100,000 dwt accounted for 66 percent of iron ore transported. In both steel plants and shipyards, not only did these changes require larger physical plants, they also required new skills, more complex organization, and computerization (O'Brien 1992; JAMRI 1991). Basic oxygen furnaces (BOF) and continuous casting (CC) technologies reduced material and labor costs and enhanced steel quality while expanding the minimum efficient scale. Lighter, stronger steels and new forms of propulsion enhanced ship capacity. Because both industries were expanding rapidly in Japan, new plants could adopt the new technologies and new organizational forms at far less cost; they also had far greater incentive to do so than the established industries in the United States did.

Transport innovations and economies of scale made possible rapid changes in the world-system hierarchy and division of labor. Japanese-led transport innovations and port construction in Brazil and in Australia, for example, led both countries to their preeminence in mineral exports. Similarly, many of the dramatic shifts in the location of industrial production in the post–World War II period would simply have been impossible without transport innovations that could reduce, or at least prevent a dramatic increase in, raw materials costs.

Japan initially had small domestic resources of a few raw materials and had rapidly exhausted these resources after World War II; moreover, Japan was far from most raw materials–producing regions after World War II, except for those controlled by China and the Soviet Union and therefore geopolitically inaccessible. Japanese industrial growth thus depended on efficient long-distance transport systems scaled for large volumes of raw materials. Japan responded by taking the global lead in developing, building, operating, and utilizing larger and more efficient ships, promoting the development of large ports, promoting the exploitation of scale-accessible deposits, and locating domestic heavy industry around large ports to allow raw materials imports to be unloaded directly to consuming plants, eliminating the need for internal transport of these large volumes of raw materials.

The complex coordination between firms within and across sectors and between firms, sectoral associations, and the state is essential to the management of these transport systems and of the raw materials industries they depend on and sustain. It is in this sense that even though raw materials costs have tended over time to constitute a smaller share of the production costs in developed economies, the technical, political, and organizational solutions to the problem of securing cheap and stable access to raw materials continues to stimulate the domestic economy's integration with an ever-widening system of external economies that defines both the conditions of ascent in the world-system and the restructuring of world markets that ascendant economies must achieve. Even more than was the case in the dramatic increase in world transport complexity and scale opportunities presented by the twin development of rail and steamship integration around the world under British hegemony, the key to taking advantage of the increasing economies of scale made available by the technological advances in ship construction in the post–World War II era was the careful matching of all stages of the transport system in order to minimize its total costs (Kendall 1972; Jansson and Shneerson 1982; Garrod and Miklius 1985). Tailoring transport systems from the mine to the consumer to take advantage of these economies of scale dramatically reduced the cost of importing raw materials on a per-ton basis.

However, this careful matching of the various components of raw materials transport systems carries important risks for the sellers of raw materials. Tailoring mines, inland transport systems, and port facilities to those of their customers' ships, importing ports, and processing plants increases the economies of scale but reduces the number of potential buyers of their raw materials, which places them in a weaker bargaining position with buyers. Selling to other potential customers whose shipping and importing facilities do not match the characteristics of the exporter may result in increased storage costs, underutilization of inland transport systems and ports (with resulting higher per-ton operating costs), and lower prices because these potential customers require lower prices to offset higher ocean transport costs per ton. The careful tailoring of mines and export transport infrastructure in many raw materials industries in Australia, Brazil, and Canada, among others, to Japan's shipping and import infrastructure gave Japanese firms important advantages in bargaining over the prices of raw materials purchased from these nations and in many cases established the bases for other forms of investment, including joint ventures (JV) and long-term contracts (LTC).

The risks for the resource-exporting nation of tailoring transport facilities to the organization created by a single rising economy are exacerbated by two factors: one, the secular tendency for transport facilities to require ever-larger amounts of inflexibly sunk capital; and two, the tendency of the importing nation to devolve as much of the cost of infrastructural development on the exporting economy as possible. Japan's need to do this was heightened in part by a severe shortage of foreign reserves at the beginning of her industrial development, but the lessons learned during that period have enabled Japanese firms and the Japanese state to keep their capital risk relatively low.

The contrast between the 1970s Japanese strategies of JVs and other forms of shared equity and host state ownership of transport infrastructure and 1950s and 1960s U.S. strategies of direct foreign investment with firm ownership of transport infrastructure, though, reveals more than mere variations between nations or eras. By the 1970s, decades of self-managed Import Substitution Industrialization, together with a growing peripheral nationalism generated within modern international developmental ideologies, had fostered a far greater capacity and disposition in a number of resource-rich countries to assume substantial equity and managerial control in mineral extraction, processing, and transport (see Becker 1983; Shafer 1994; Girvan 1990; Bosson and Varon 1977).

Japan used this new situation to her own advantage and in the process changed not just the organization of minerals markets and transport systems but also the rules—or the regimes—within which states and firms in exporting countries must make their decisions (Bunker and O'Hearn 1992). In this sense, while the Japanese led the transport revolution that changed raw materials markets and costs, they largely depended on exporters' acquiescence and active participation. Japan's success in securing peripheral cooperation enhanced domestic economic growth while dramatically increasing global inequalities; in their combined external and internal dynamics and effects, raw materials access and transport thus constituted the national economy's generative sector more directly than at any time since they undergirded Holland's ascent three hundred years earlier.

A variety of factors combined to produce an increase in the size of ships carrying other major raw materials, especially iron ore and coal. These included the development of ever-larger oil tankers; the increasing size and experience of shipyards, especially in Japan, in building larger ships; the growing volume of other raw materials trades; the recognition that economies of scale also existed for other raw materials shipping; and the increasing average distance to adequate resource supplies. In the mid-1960s these trends combined to increase significantly both the average and the maximum sizes of dry-bulk carriers. The number of dry-bulk carriers (ships specially built to carry minerals and grains) increased by a factor of ten between 1961 and 1992, from 471 to 4,846, while the average size of each dry-bulk carrier increased from 18,495 to 44,552 dwt. The cumulative result was a dramatic increase in the total available tonnage for shipping minerals and grains—from 8.7 to 215.9 million dwt—during these three decades (Fearnleys, various years; UNCTAD 1969), which was commensurate with the tremendous increase in minerals transport during the period.

Japanese government and firm efforts to reduce transport costs were of particular importance given the tremendous distances which Japanese raw materials imports had to travel. Japan had a tremendous disadvantage relative to Europe and North America in iron ore and coal, the second and third most important raw materials traded in terms of volume (behind petroleum), with Japan's shipping distance for iron ore averaging 6,140 nautical miles in comparison with 4,300 miles for Europe and 2,970 miles for the United States, and the Japanese shipping distance for coal imports averaged 6,240 miles in comparison with Europe's 3,020 miles and North America's almost entirely continental supply (Drewry 1978: 1). This differential created a tremendous incentive for the Japanese government and firms to reduce transport costs, since without such reductions these long distances would have made Japan uncompetitive in raw materials–based industries.

Following a path laid out by Holland, Great Britain, and the United States early in their respective ascents, the Japanese government targeted shipbuilding as a critical sector for its reconstruction after World War II. It soon became a generative sector in Japan's rise to trade dominance. We will examine the strategies and the state-sector-firm financing collaborations that drove Japan's rise as a shipbuilding and shipping nation in the following section.

Transport as a Japanese Raw Materials Access and Development Strategy

Transport played a variety of roles in Japan's post–World War II raw materials access strategy and in its broader economic development strategy. These strategies in turn critically supported Japan's rise to trade dominance during this period. By 1984, Japan accounted for 17 percent of total world seaborne imports in terms of volume because of its huge volume of raw materials imports, making Japan by far the world's major importing nation. Because they consisted of industrial products of much lower volume and much higher value, Japanese exports were only 3 percent of world total exports in terms of volume in 1984 (Stopford 1988: 141)—but Japan's share of world exports in terms of value is much greater, 8.9 percent in 1984 (Chida and Davies 1990: 184). Japan's rapid economic growth both drove and depended on the rapid growth of raw materials imports, up to a total of 491.7 million tons by 1986, including 115.2 million tons of iron ore and concentrates, 91.4 million tons of coal, 157.4 million tons of crude oil, and 59 million tons of refined petroleum products (USBM 1987: 509–14).

Transport planning and subsidies specifically aimed to support Japan's tremendous expansion of raw materials imports at competitive cost levels. Petroleum, iron ore, and coal have been the most important imports in terms of volume, although bauxite, alumina, aluminum, copper concentrates, liquefied natural gas, and a host of other minerals were also imported in increasing volumes during the period. The state and firms collaborated in research and development on the construction of larger petroleum tankers and bulk carriers in order to capture economies of scale in construction and operation. This required them to develop extraordinarily large shipyards. The Japanese shipbuilding industry became the world's leading builder and exporter of ships, as well as a major consumer of steel. The Japanese state enhanced the generative effects of shipbuilding and shipping by targeting steel to lead the Japanese economy, both as exported directly or incorporated into manufactured goods and as incorporated into domestic construction of industrial, transport, and business infrastructure.

The ownership and operation of such large ships by Japanese shipping firms associated with major industrial groups, including those that owned steel mills, allowed these firms to free themselves from their dependence on foreign shipping for imports and exports, giving them greater control over their production costs and their costs of supplying goods to foreign markets. Japanese raw materials consumers' control over ocean shipping meant that they captured any reductions in transport costs caused by technological improvements or changes in world shipping market conditions, even if these improvements were funded by the raw materials exporters.

The construction of large-scale port and railroad infrastructures at the expense of raw materials exporters (firms and/or states) allowed the efficient use of these large ships at no cost to Japanese firms or the Japanese state. Domestically, on the other hand, the Japanese firms and state did finance the construction of Maritime Industrial Development Areas (MIDAs) in Japanese ports. This massive rebuilding of the urban-industrial environment eliminated the need for the internal transshipment of raw materials imports, reducing raw materials import costs and allowing the government to redirect industrial development to new areas of the nation. The MIDAs constituted the link between the domestic and the international components of Japan's generative sectors.

Japan's geographical and geopolitical situation alone, of course, could not account for the complex and ambitious projects undertaken to resolve the diseconomies of space. Ambitions for trade dominance are formed in economic and political culture, not in geography. We saw in the earlier cases of national ascent to trade dominance, though, that political and economic culture were molded at the intersection of preexisting cultural forms with local configurations of matter and space mediated by available technology. Similarly, Japanese political culture had developed toward national industrial development within the context of the island's limited domestic raw materials and relative isolation from world sources and markets.

State coordination with private firms to promote transport technologies and infrastructure has been a major component of economic development strategy in Japan since the Meiji Restoration. The state negotiated the location and construction of ports with local warlords in exchange for their submission to national authority. It overcame dependence on foreign-owned shipping by subsidizing and protecting Japanese-owned coastal and international shipping companies competing with the foreign-owned lines, a policy that has continued in a variety of forms until the present day. The high cost and foreign exchange drain of importing ships built in Europe also led the Japanese government to provide support for Japanese shipbuilders, another policy that has been maintained ever since. The state licensed, and then subsidized, transport within the great business groups—*zaibatsu* until the World War II defeat, afterward *keiretsu,* the reconstituted form that emerged under U.S. postwar tutelage. All of the policies supporting ports and shipping created *zaibatsu/ keiretsu* expectations and acceptance of government intervention and of state-firm collaboration.

This government direction and subsidization reflects the importance of transport in state efforts for raw materials access and for economic development more generally. Japanese government policies during the 1950s favored the development of a Japanese-owned liner fleet to carry Japan's light industrial exports. These policies shifted in the early 1960s to emphasize the construction of oil tankers and ore carriers to transport Japan's rapidly growing raw materials imports and then to build containerships at the end of the 1960s (Goto 1984: 8). This government direction was critical to guaranteeing large volumes of raw materials at low cost to Japanese firms.

As we saw in the cases of Holland, Britain, and New England, though, efficient access to raw materials through cheap, efficient shipbuilding requires cheap, efficient access to the raw materials required for shipbuilding. After World War II Japan had neither, and its success depended on establishing synergies between access and construction that would cheapen both. The military procurement strategy that Britain adopted in similar circumstances had served Japan badly before World War II and was impossible after the war.

The Japanese solved this problem through close economic planning, collaboration between states and firms in a situation where nearly all sectors of the economy were dependent on resources, and close coordination with (and occasional defiance of) the United States. Shipbuilding and the steel industry constituted the keystones of Japan's planned domestic development and of its international access to cheap, stable raw materials supplies. The two industries sustained each other in critical ways. An efficient shipping sector expanded the volumes and lowered the costs of the iron and coal essential to Japan's successful competition in the world market; cheap, high-quality steel enabled economical construction of the huge ships needed to import these critical raw materials. The Japanese strategy for developing a competitive steel industry was based on promoting plants that maximized scale economies and then assuring that they ran at full capacity. At its apogee, the shipbuilding industry absorbed 35 percent of steel output. It also fostered a number of steel-consuming ancillary industries that eventually became autonomous (Yamamoto 1980).

In addition to government support through financing, government support for research and development on ship construction also played a critical role in the post–World War II era. The single most important aspect of government support for the technological development of shipbuilding in Japan was the agreement signed between the Japanese government and the U.S. shipping firm National Bulk Carriers (NBC) in 1951. The booming ship construction market had led NBC (owned by U.S. shipping magnate D. K. Ludwig) to search for a former naval shipyard that could be utilized for civilian shipbuilding, especially for large-scale petroleum tankers (Todd 1991: 13). NBC investigated a number of former naval shipyards in Germany and Japan and selected the Kure naval shipyard in Japan. The terms of the agreement between NBC and the Japanese government on the ten-year lease were that "while NBC was to construct ships for its own purposes as and how it wished, as much Japanese steel as possible was to be utilized. Of even greater long-term significance was the insistence that all types of Japanese shipbuilders and engineers were to have free access to the establishment and were to be permitted to examine all aspects of its building system" (Chida and Davies 1990: 112). At this new shipyard, the technological and organizational systems that made it possible for Japan to become the world's leading shipbuilder were developed under a unique set of lease terms intended to transfer advanced technology to Japan and to train Japanese engineers and managers at no cost to Japanese firms or to the Japanese government.

Beginning in the early 1960s, other Japanese shipyards began introducing a number of technological innovations of their own based on the experience and technology introduced to Japan by NBC, including replacement of labor with machines in the hull construction and fitting-out departments. Ship design was also rationalized and production control methods were improved. The introduction of "section" or "block" building and welding allowed the development of larger shipyards with more efficient layouts that permitted the construction of larger ships. All of these improvements increased labor productivity and reduced the amount of labor required to build each ship, increasing the competitiveness of Japanese shipbuilding relative to the more labor-intensive methods that remained in use much longer in other nations (Chida and Davies 1990: 91–92). Japan remains a world leader in shipbuilding technology, particularly in the reduction of labor requirements per ship through the introduction of capital-intensive mechanization and improved management.

Technological and organizational improvements in Japanese shipyards during the 1950s and especially the 1960s gave Japanese "shipbuilders sufficient economies of scale that they could lead the world in the new technology" of building extremely large oil tankers that were in increasing demand after the closing of the Suez Canal and as the distance between oil-extracting regions and their customers increased (Chida and Davies 1990: 98–99). The rise of general freight rates while the Suez Canal was closed included an increase in the cost of shipping other bulk materials, including iron ore. In the case of iron ore, the share of ocean freight costs rose to more than half of the CIF ("cash, insurance, freight," representing the major costs paid for by the seller) price of iron ore imported to Japan. The rising costs strongly motivated the creation of a large fleet of Japanese ore carriers.

The tremendous increase in the size and number of Japanese-owned bulk carriers starting in the 1960s resulted from state policies that simultaneously linked, subsidized, and regulated the steel and shipbuilding industries (Chida and Davies 1990: 119). As the result of these technological innovations and the tremendous demand in Japan for bulk shipping, "Japan then emerged as the world leader in the production of very large vessels and subsequently dominated the market for oil tankers and ore-carriers. The timing of this development was particularly fortuitous for Japan as it occurred just as a boom began which required substantial numbers of large tankers" (Chida and Davies 1990: 133). The Japanese state also provided financing on concessionary terms for the export of Japanese-built ships through the Export-Import Bank of Japan.

Additionally, state funding of technological modernization of the Japanese steel industry in the 1950s led to major cost reductions of the steel plate used in shipbuilding and thus to making Japanese ship prices much more competitive (Chida and Davies 1990: 108–9). Japanese steel firms initially intended to build their own fleets of ore carriers, but the Japanese state's refusal to finance shipbuilding and purchase by firms other than the major shipping lines forced the Japanese steel firms to invest in shipping firms if they wanted to secure at least partial ownership of bulk shipping. This policy kept shipping concentrated in the hands of a small group of large firms whose development of new technologies the state could both subsidize and direct.

With this support from the state, by 1956 the Japanese shipbuilding industry had become the world's largest. It produced about half of total world output from the mid-1960s into the 1990s. Shipbuilding was also Japan's most important export industry between 1956 and 1960 (when the steel industry surpassed it in exports) and remained one of Japan's three major export industries well into the 1980s (Chida and Davies 1990: 106). At the same time, shipping efficiencies lowered the costs of the raw materials most critical for competitive shipbuilding. By 1970, Japanese raw materials costs in steelmaking were marginally lower than in the United States (Patrick and Sato 1981: 12), an extraordinary achievement considering Japan's locational disadvantages. In these ways, shipbuilding and steelmaking provided the base for Japanese trade dominance. In the following section, we examine how the national state's collaboration with heavy industrial firms and with finance not only supported the globalizing efficiencies of these sectors but how their relations around these industries generated far more general developmental dynamics in the national economy.

How a Defeated and Dependent Nation Rose to Trade Dominance

The Allied victory over Japan in World War II and the subsequent division of global commercial and military systems between the Soviet Bloc and the countries allied with the United States closed off Japan's proximate sources of the raw materials most voluminously used for industrial production. The desire to contain Soviet expansion, however, motivated strong U.S. financial, technical, and diplomatic support for rapid Japanese industrial recovery and expansion. This included, first, direct supply of U.S. raw materials and then, in order to save scarce dollars, financial and diplomatic assistance in arranging alternative sources.

The push to industrialize, started by the Meiji regime and intensified through a series of regional and world wars, seriously depleted Japanese sources of iron and coal. Indeed, Japan's aggressive strategies to procure access to proximate foreign sources of matter and energy prompted the U.S.-British embargos on oil and other raw materials that eventually provoked Japan to attack Malaya and Pearl Harbor (J. Marshall 1995). Thus, U.S. proscription of commerce with China, North Korea, and the Soviet Union after the war, together with lingering fear and resentment of Japan in its other former raw materials–supplying colonies, simply compounded the socially and materially destructive consequences of earlier raw materials procurement policies. Iron and coal in the volumes needed for internationally competitive economies of scale were thus available only at considerable distance, expense, and diplomatic effort.

Initial Allied plans for Japan's postwar reconstruction imposed strict limits on the Japanese steelmaking, shipping, and shipbuilding industries. Rapidly rising tensions with the Soviets in the late 1940s moved the United States to end its restrictions on Japan's shipping, shipbuilding, and steelmaking industries in order to reconstruct its former foe as a geopolitical bulwark against communism in Asia (Chida and Davies 1990: 62–65).

The shift in U.S. policy coincided with and complemented Japanese state programs to reconstruct the national economy (Bunker and Ciccantell, forthcoming). The Priority Production System regulation of December 1946, for example, identified several key industries as leading sectors to be subsidized and promoted. The proclamations announcing the creation of the system strongly emphasized coal and steel. The Programmed Shipbuilding Scheme of 1947 supported the growth of both shipping and shipbuilding. This program aimed to speed the construction of shipping to ease the severe postwar shortages. The state provided low-cost financing to private shipping firms, initially through the Rehabilitation Finance Bank and, after 1953, through the Development Bank of Japan.

The Programmed Shipbuilding Scheme based its selection of shipping firms to receive these loans on government estimates of the need for various types of new ships. The shipping firms then used these loans, supplemented by loans from commercial banks, to order ships from Japanese shipyards. The Japanese government thus subsidized and regulated shipping firms directly in order to develop a domestic fleet to supply the transportation required for economic reconstruction. Indirectly, these subsidies supported Japanese shipyards and steel mills (Chida and Davies 1990: 89–90). U.S. demand for ships during the Korean War dramatically increased freight rates and greatly strengthened Japanese ship owners and shipyards (Chida and Davies 1990: 66–86).

The diseconomies of space that confronted Japan's need for expanded supplies of raw materials, enhanced by U.S. proscription of proximate sources, impelled the Japanese to devise new forms of closely coupled state-firm and inter-firm collaboration in order to implement historically unprecedented advances in economies of scale in both production and transport. State agencies, including government banks and the Ministry of International Trade and Industry (MITI), encouraged interfirm collaboration and information-sharing for adopting and improving optimal existing technologies, particularly in heavy industry and transport, and to coordinate their raw materials–procurement strategies and contracts. Expanded transport capacity so enhanced Japanese firms' bargaining capacity that they were able to combine enticements and pressures that convinced numerous firms and states in resource-rich nations to invest in the largest iron and coal mines in history and then to sell the minerals extracted at some of the lowest prices in history (Bunker and Ciccantell 1995a, b, 2003a, b, forthcoming; Ciccantell and Bunker 2002).

The investments needed to develop the technologies and construct the infrastructure that created these economies of scale exceeded the capital available to particular firms through single financial agencies. They therefore required unprecedented expansions in the size and power of financial institutions, supported by unprecedented levels of state capacity and authority to coordinate its closely coupled collaborations with national firms, sectors, and finance. The propensity of the Japanese public to high rates of saving and their acceptance of the very low interest rates paid by the state-controlled Postal Savings Bank, combined with sectorally targeted loans from the World Bank and investments by U.S. firms in strong currencies that could be favorably converted into weak yen to cover domestic expenses, provided the liquid capital that state-controlled banks needed to provide low-interest loans for selected firms and associations in targeted sectors. Central banks enhanced these loans by allowing inflated value estimates on the real estate used as collateral. Urban land prices soared as a result. These practices enabled the banks to supply the huge capital needed to create the economies of scale required to make Japanese steel and Japanese ships cheap enough to support Japanese competition in global markets.

The technological and organizational strength that these state-firm-finance collaborations gave the national economy, together with the extraordinarily increased economies of scale in production and transport they achieved, enabled Japan to globalize raw materials markets and transport in ways that led to trade dominance in these and other sectors in less than two decades. Complex state-supervised financial structures, combined with MITI regulation of the vast capital sunk in heavy industrial and transport infrastructures on the scale required to overcome the diseconomies of space the nation faced, made it possible to stimulate hugely expanded *international* flows of cheap raw materials to drive rapid *domestic* economic growth. Japanese state agencies coordinated multi-firm negotiations and financial packages to encourage states and firms in resource-rich nations to invest in mines and transport systems that would extract and deliver minerals to huge new ports on a scale compatible with the huge dry-bulk carriers the Japanese were building; the iron mine, railroad, and port for Carajás in the Amazon is only one prominent example. The Japanese thus achieved global excess capacity in most major minerals. This lowered raw materials prices and left indebted extractive firms and states susceptible to proposals to increase their subsidies to transport and to reduce their mineral rents and prices even more.

Low raw materials and transport costs contributed significantly to Japan's economic boom. The dramatic success of state promotion and regulation of raw materials contracts, shipbuilding, and heavy industry from the 1960s through the 1980s enhanced the already high degrees of state control over banks and private firms. The growing power and legitimacy of state control over business, though, together with Japan's continued subordination to U.S. military and geopolitical goals and needs and her continued dependence on U.S. product and financial markets, left her more vulnerable to economic crisis and policy change in the established hegemon than other rising national economies had ever been. Britain benefited from Dutch investment and from Dutch markets during its rise to trade dominance; similarly, the United States drew on British finance while profiting from British demand for raw materials. Neither of the earlier ascendant economies, though, were as directly subordinate to, nor as integrally entwined with, the established hegemon's geopolitical agenda. Indeed, both fought wars against the established hegemon during their rise.

Successful management of the iron-, coal-, and steel-based generative sector had been based on, and had intensified, both external dependence on the United States and the internal authority of the central state. These two conditions combined to leave the Japanese state vulnerable to direct U.S. intervention in commercial and fiscal policy. During the 1980s, the fiscal imbalances occasioned by President Reagan's transformation of the United States from global creditor to global debtor and the subsequent pressure from the U.S. State Department for Japan to strengthen the yen against the dollar (Murphy 1996) and to open Japanese capital markets to U.S. participation (Gao 2001) destabilized and devalued the financial and productive systems that had made Japan's rapid economic growth possible. The Japanese government and its tightly controlled banks used their moral authority and legitimacy with private financial institutions to support the valorization of the yen to which the 1985 Plaza Accords committed them. The resulting strength of the yen against the dollar encouraged Japanese firms and banks to invest massively in U.S. real estate, but the U.S. economic crisis of the late 1980s and early 1990s devalued these holdings dramatically. U.S. leverage over Japan's fiscal policies combined with the control over finance that the Japanese state had achieved during its active coordination and subsidization of the rapidly expanding steel and shipping sectors thus contributed importantly to the Japanese crisis and recession that started in the 1990s.

In brief, the same closely coupled coordination with firms, sectors, and finance that enabled the Japanese state to organize economic expansion enabled the United States to pressure it to adopt financial policies that favored the United States rather than itself. The resulting crisis in the Japanese economy shows that its earlier success was very much conditioned by U.S. support in pursuit of U.S. geopolitical goals. That support enabled Japan to globalize markets for low volume bulk raw materials, to catalyze the transport systems that led to this culminating stage of globalization, and in the process devolve much of the cost of globalization to its own raw materials peripheries.

The crisis in Japan repercussed badly on its indebted raw materials suppliers. Many had yen debts to Japanese and multilateral banks, as well as supplier credits to Japanese firms. These became more costly to service as the yen grew stronger. At the same time, because international contracts for sales of most metals are specified in U.S. dollars, the value of their raw materials exports sank as the dollar weakened against the yen. These problems were compounded by Japanese renegotiations or violations of the LTCs as their economic downturn slowed raw materials consumption. Japan's own rapid political and economic development and prosperity reduced the economic, political, and ecological integrity of its raw materials suppliers.

Technological Innovation, Financial Strategy, and Bargaining Skills

Japan's rapid rise and subsequent stagnation embody the most recent, and therefore the most extreme, of the historically reiterated intensifications of matter and expansions in space brought about by technically expanded economies of scale. The commercial and financial relations required to sustain this unprecedented materio-spatial intensification and expansion generated proportional increases in global inequality. The Japanese drive to trade dominance dramatically enlarged the spatial separation of extraction from production. In addition to enlarging the technological and financial scale requisite to globally competitive commerce in bulky raw materials, the increased distances tremendously exacerbated the political, economic, social, and ecological inequalities between extractive and productive societies. In order to overcome the huge diseconomies of space imposed by the economies of scale that a nation needed to ascend in the world-system of the late twentieth century, Japanese strategies for access to raw materials could only succeed by reducing rents and devolving transport costs onto suppliers. Firms, sectoral associations, banks, and state agencies learned together the technological, organizational, financial, and diplomatic skills essential to Japan's radical transformation and spatial extension of global raw materials and transport markets and to the extraordinarily rapid rise to trade dominance that its control of these markets made possible.

Until the 1960s, U.S.-led FDI in mineral extraction created oligopolistic control of major minerals markets. The oligopolies had managed to lag supply sufficiently behind demand to keep minerals prices high (Barham, Bunker, and O'Hearn 1994). This kept significant profits near the mouth of the mine, thus enabling host nations to exact significant rents, even if profits, and the stream of linked benefits in transforming matter and energy into commodities, were repatriated to the core. By the mid-1960s—and then much more powerfully in the 1970s as OPEC showed how international collaboration could enhance resource rents—the loss of these repatriated profits spawned growing resource nationalism in the raw materials—exporting periphery.

The Japanese learned how to exploit this nationalist resentment. In the 1960s, Japanese firms, sectoral associations, banks, and state agencies devised and aggressively promoted new forms of international investment (NFI; cf. Bomsel et al. 1990), such as LTCs and JVs, with local and national states and with multinational and local mining companies. Efficiency in shipping, and the closely coupled collaboration that state agencies, financial institutions, and sectoral associations had developed to achieve it, enabled the Japanese state and firms to induce multiple resource-rich nations to invest in huge mines and in dedicated rail and port systems.

The Japanese became increasingly adept at reducing their capital exposure in these ventures (Bunker 1994a). They used these NFI, together with complex, highly detailed, and very optimistic projections of their stimulating effects on local economies, to induce peripheral states and firms to invest in mines and transport infrastructure. The Portuguese-language publication of the report contracted by a consortium of Japanese sectoral associations and state agencies to project the benefits that they claimed the Carajás iron mine and the Albras aluminum smelter would bring to the Amazon's economy, for example, ran to sixteen volumes. Distributed to key local state planning agencies, it was expensively bound and printed on high-quality glossy paper with elaborate graphs and maps of the linkages and spread effects anticipated for each project.

By the 1980s, similar strategic Japanese inducements, carefully tailored to the geographical particularities, political systems, nationalist ideologies, and industrial aspirations of Australia, Canada, Venezuela, Indonesia, Chile, Papua New Guinea, and Mexico, among others, had created significant excess capacity in the minerals that most interested the Japanese—iron, coal, aluminum, and copper—with little and sometimes no capital risk to Japan. They had also swelled the foreign debts of the resource-exporting firms and nations. Japanese firms and sectoral associations used their suppliers' overcapacity, debt dependence, and market competition to bargain down resource prices and rents. In many cases, they reduced their own equity participation in projects, as they did in Brazil, Indonesia, and Venezuela (Bunker 1994a), or reduced the mineral take to which their LTCs committed them, as they did in Australia and Canada (Ciccantell and Bunker 2002; Bunker and Ciccantell 2003a, b).

A similar process occurred in port facilities. The close articulation between the reorganization of the Japanese environment—which cheapened raw materials access by relocating heavy industry near large ports—and the reorganization of world transport systems—which accommodated the transport scale required for the success of Japanese heavy industry—becomes evident in an analysis of port construction in the regions importing and exporting the most coal and iron. In 1980, there were three iron-exporting ports in the world capable of handling ships of 250,000-dwt capacity. By 1989, there were nine, of which three, all in Latin America, could handle more than 300,000 dwt. There are no berths for iron ore shipments this large in the United States, only four in the European Union, and twelve in Japan. Coal is generally shipped from smaller harbors than iron ore, but here the differences are equally dramatic. In 1980, the largest exporting facilities for coal were two harbors in Canada capable of handling 110,000 dwt. By 1989, there were eight coal-exporting harbors in the world capable of handling 130,000 dwt or more. There are seven berths in the European Union large enough to unload a ship this size; in Japan there are thirty (calculated from Tex Report 1994a, b). As we remember how crucial ship and port size are to reducing transport costs, we can see how great were the trade advantages Japan captured by engineering the spatial shape and distribution of increased scale in the global built environment.

They thus first encouraged multiple peripheral suppliers to assume most or all of the infrastructural and operational costs of mines and dedicated transport systems scaled to the size and distance of Japanese smelters and then violated the terms of LTCs that peripheral states and firms had used to secure financing for building the mines and transport systems that they had themselves demanded as a condition of the LTC. In many cases, the peripherally owned mines had to continue operating in order to amortize the debts they had assumed, even when prices fell below their costs. They thus contributed to even greater excess world supply and to even lower prices (for more detailed accounts, see Bunker 1994a; Ciccantell and Bunker 2002; Bunker and Ciccantell, forthcoming).

Unequal distribution of the benefits provided by nature has been the aim of all core strategies for trade dominance. The arrogant European phrase "free gift of nature" justifies core pressures on extractive economies to reduce rents and prices of natural resources and legitimates the presumed superiority of value created by labor over value created by nature (Coronil 1997). It also ignores the social and ecological damage that large-scale extraction imposes on natural, as well as on other social, production systems. Japan's innovations in transport and in state-firm collaboration, together with elegantly packaged, complexly enumerated analytic projections of how supplying these free gifts of nature to Japanese industry would accelerate local economic development, enhanced the ability of Japanese firms and sectoral associations to reduce rents and devolve infrastructural costs on a far greater scale than any previous rising economy had ever done.

Japanese success in breaking FDI-based oligopolies by inducing multiple foreign firms and states to risk their own capital and debt in large-scale mining and transport systems created excess capacity in key world mineral markets. By playing their separately financed suppliers off against each other, Japan managed to reduce the rents and profits in extraction, to devolve the costs of overland shipment and of port construction and operation to others, and thus to reduce the price and the landed cost of raw materials. By reducing their equity and capital exposure in mines and in overland transport systems, Japanese firms could shift profits further down the commodity chain, that is, away from the mine mouth and into industrial transformation. In order to improve their competitiveness and increase their market share in international trade, the Japanese state encouraged national firms and banks to sacrifice some of these profits to lower the prices of exported commodities. The result was Japan's rapid ascent to dominance in world markets for manufactured commodities and a significant presence in world capital goods markets.

Like earlier trade-dominant nations, Japan soon added high-value-added goods to its list of competitive exports, and, like its predecessors, it did this from the strong base it had already achieved in heavy industries and bulk transport. It then instituted flexible just-in-time specialization systems to reduce warehousing costs and to enhance responsiveness to buyer demand, but here too its ability to do so depended on strong throughput in the more basic industries.

Japan, even more than earlier trade-dominant nations, used two interdependent means—expanded economies of scale in the material processes of production and transport, on the one hand, and innovative social and financial regimes of coercion, contract, and persuasion, on the other, to pry cheap and stable access to raw materials out of the globe's social and natural systems. The enormous advances in transport efficiencies that Japan developed in the 1960s, supported by the economies of scale and quality that Japan achieved in steelmaking, globalized its access to raw materials. Privileged cheap access to raw materials remained as important as ever to developing a competitive national transport system, but Japanese achievement of global economies of scale in transport broke the dependence on geographical proximity to sources that had constrained earlier rises to trade dominance. Except for the facility that Japan's extraordinarily deep and indented shoreline provided for constructing multiple deep-water ports to service a dense chain of industrial cities, its comparative advantage in the cost of landed raw materials was technologically and socially, rather than naturally, produced.

The incentives to produce this technology, however, were rooted in the intersection of Japan's geographical and geopolitical situation at the end of World War II. Japan's distance from its raw materials sources and from the commodity markets it serviced made it even more dependent than earlier trade-dominant nations on cheap raw materials and cheap transport. The strategies the Japanese devised to satisfy these twin needs globalized raw materials markets, but the investments required to make global sourcing economically competitive were beyond Japan's capacity. Japanese firms and state thus devised strategies to devolve a significant share of the costs of resolving the diseconomies of space on their raw materials suppliers. At the same time, they devised innovative financial institutions and policies to combine and coordinate the separate capitals controlled by disparate firms, banks, and state agencies, with capitals in multilateral banks and foreign states and firms. Their success in these paired strategies increased global inequalities even more dramatically than earlier trade-dominant nations had done.

Japanese Dominance of Steel and Shipbuilding in Comparative Perspective

There are striking parallels between the Japanese experience of steel and shipbuilding and the Dutch experience of wood and shipbuilding three centuries earlier. Both nations achieved the world's cheapest means of transporting the raw materials used most voluminously—in ship construction as well as in industry and infrastructure generally. In both cases, cheap raw materials enabled technological innovations and economies of scale in construction that in turn supported imports of greater volumes of raw materials. The expanded throughput of raw materials allowed further economies of scale in shipping, handling, and processing that created opportunities for advances in technical and organizational efficiencies.

The high levels of growth and throughput achieved through the close coordination of the shipbuilding, shipping, and raw materials industries most critical to shipbuilding supported Japanese economic ascendancy as directly as they had Dutch economic ascendancy. The efficiency and speed of bulk transport had increased dramatically over the intervening three centuries, though, as had the engineering and construction capabilities for building environments, so that the physical proximity and topographical advantages that the Rhine, the Elbe, and the Weser had provided to Holland found no direct counterpart in the Japanese case. Instead, the Japanese took advantage of their extensive and indented coastline—and of the willingness of key raw materials–exporting nations to build their own ports and railroads—to build a tightly integrated, highly efficient environment that connected foreign mines to foreign ports to Japanese ports to smelters and then to downstream fabricators. State-sector-firm collaboration thus combined interdependent heavy industries in ways that created comparative advantages in bulk handling quite similar to those enjoyed by the Dutch, but on a much larger scale. Whether socially constructed or naturally encountered, an environment favorable to the cheap bulk transport of the raw materials they used most voluminously was critical to both Japan and Holland. The near-total dependence on imported raw materials obviated the sectoral strife and accommodations that the British and American national states had been forced to make. In both Japan and Holland, this facilitated a more precise implementation of materio-spatial logic in the articulation of their policies and strategies regarding raw materials access, technological development, industrial location, and state subsidies and targeted loans for key sectors.

For the Japanese economy, as in the earlier Dutch economy, planning and coordinated implementation by industry and the state clearly followed the same materio-spatial logic (O'Brien 1992; Yonezawa 1988). Both the shipbuilding and the steel industries enjoyed far greater subsidies than were provided to other sectors, and only shipbuilding received interest subsidies on loans (Ogura and Yoshino 1988). Both industries were critically involved in and affected by wider Japanese policies to assure cheap and stable access to raw materials, including direct investments, loans, and LTCs aimed at stimulating the development of compatible overland transport and port facilities in various raw materials–exporting countries. Just as Dutch and New England shipbuilding intersected dynamically with a vibrant timber economy that facilitated their world dominance in shipping, so the Japanese shipping industry intersected dynamically with the domestic steel industry, with each reinforcing the strength of the other. Because building a large ship requires the complex coordination of multiple specialized processes in ways that often must be tailored to a particular buyer's transport needs, shipbuilding gave rise to a series of ancillary industries. Some of the skills and technologies these industries developed could be adapted to other applications and products; these industries eventually became autonomous and then dynamic in their own right (Yamamoto 1980).

Steelmaking in turn "became the archetype of Japan's post-war success" (O'Brien 1992: 128) and a proving ground for effective state intervention and state-sector-firm coordination in economic planning. MITI used its central role in securing finance and cheap raw materials to enable and coordinate multiple firms' adoption of new technologies and the construction of huge port and industrial complexes large enough to contain the huge integrated steel mills in close and convenient proximity to the industries that consumed their output downstream. It thus achieved enough power over the individual steel mills to constrain their separate impulses to expand their market share enough to create excess capacity. Though it had to compromise in its negotiations with the most powerful firms, MITI achieved a general acceptance of its goal to build only smelters that could operate at full capacity even with soft demand, using imports to make up the shortfall when demand increased. This closely regulated but intimately supported steel industry was able to provide Japanese manufacturers with high-quality, low-cost steel that became a key element in their ability to penetrate global commodity markets.

As we have seen with other national economies rising to trade dominance, the huge capital costs and the massive rebuilding of the environment required to coordinate access to and economical transport of the most voluminously used raw materials generated not only new, larger-scaled technologies but also larger financial institutions and new, more complex, inclusive, and powerful interactions between state, firms, key sectors, and finance. Once again, then, the interactions between the most voluminously used raw materials and the transport systems that supply them become the generative sectors that drive national rises to trade dominance and in the process spur major advances in globalization.

Comparing the Generative Effects of Wood and Steel Ships

Britain remained dominant in shipbuilding until World War I because the size and complexity of ships were still beyond the integrated construction techniques that were revolutionizing production in sectors that produced in smaller units (Hobsbawm 1968: 179; see also Lash and Urry 1987). British shipbuilding did stimulate the advances in metallurgy and in steam engines, but it did not introduce new industrial organizational forms, as Dutch ship-building had. This meant that, however important shipbuilding was economically, financially, and politically, it was not an enduring generative sector as it had been in Holland.

The impact of steamship building on a rising economy lasts for less time and is less stable than the impact of wood ship construction; the relatively greater scale and concentration of capital in building steamships causes violent fluctuations in shipbuilding. Iron and steel allowed for much more rapid increases in boat size than had been possible with wood. The *Britannia,* launched in 1840, had a 215-foot deck and a 34.3-foot beam, with a gross tonnage of 1,139; the *Oceanic,* launched in 1899, was over three times longer, almost twice as wide, and boasted more than fifteen times the gross tonnage. By 1914, the *Vaterland,* over four times as long and three times as wide as the *Britannia,* exceeded her gross tonnage by a factor of fifty (Berglund 1931: 35). In comparison with even the largest of wooden boats, the cost of building these vessels was enormous, as was the capital sunk in the large and specialized facilities needed to do so. Maintaining full capacity production is critical to profit when large amounts of capital are sunk in specialized and inflexible investments, but the nature of ships and shipping makes this almost impossible. The scale and inflexibility of capital sunk in shipyards creates financial pressure to maintain production levels even when demand falls, so production costs exceed sale price. Even the slow increase of capacity during periods of strong demand can create serious excess capacity when demand stabilizes or falls. Prices, profits, and employment remain very volatile in this industry.

There is one condition under which shipyards have historically expanded capacity very rapidly, with long-term detriment to profit levels. To an even greater extent than was true of wooden boats, war stimulates tremendous demand for ships, because it destroys them and other goods whose replacements they must carry, because each side needs to move large amounts of troops and material. Typically, shipyard capacity is massively expanded in wartime, and wars typically end with a tremendous excess capacity of both ships and shipyards.

Because the capital in boatyards is very inflexibly sunk, excess capacity drives prices down drastically. For example, in 1918—a war year—7,144,549 gross tons of shipping were launched. That year the price of a new 7500 gross ton steamer was 232,500 pounds sterling. In 1923—with the war over—less than a quarter of the wartime level of shipping was launched, 1,643,181 gross tons. A new 7500 gross ton steamer now cost only a quarter as much, 60,000 pounds sterling. By 1929 tonnage recovered to 2,793,210, but prices continued to fall, reaching 58,000 pounds sterling. Freight rates fell by two-thirds between 1920 and 1921, and continued down to 25 percent of their 1920 level by 1929, indicating that despite the drop in construction, there was still excess shipping capacity. The collapse was even more dramatic during the following three years.

Increases in scale and concentration in an industry with high requirements of inflexibly sunk capital and a product with low rates of replacement lead to huge fluctuations in employment and production. Cheap transport remains critical to world economic dominance, but the ascendant economy must move fairly rapidly through the period when shipbuilding itself is a leading sector.

Overall, even though steamship construction had significant effects on innovation and the organization of production in Britain, these effects were relatively less significant than shipbuilding had been on the far less industrialized Dutch economy. On the other hand, the sheer volume and diversity of goods cheaply transported over great distances in steamships and the needs and opportunities involved in articulating steamship routes with railroad lines and deeper harbors, coal depots, warehouses, and stockpiles drove the British to reorganize the economies and polities of many more social formations around the world much more deeply and widely and to integrate them much more tightly than the Dutch ever had. The greater speed of communication and financial transactions enhanced both possibility and effect. The scale and inflexibility of the infrastructure around which these new organizational forms emerged created an inertia—political, economic, and infrastructural—that reinforced and sustained British carrying, trading, and financial dominance.

The Dutch had also continued to dominate world shipping and shipbuilding long after the relative decline of their internal economy. Even in the eighteenth century, this was largely the result of the huge investment in the highly durable infrastructure required for shipbuilding and port construction. The technologies of steam and iron, the huge increases in vessel size and duration that they made possible, the distances they could economically transport low-value bulk, and the capital sunk in the physically integrated overland, port, and shipping systems combined to dramatically amplify these effects. Once in place, the large quantities of capital inflexibly sunk in shipyards create an infrastructural inertia that may be extraordinarily durable. Cafruny (1987) notes that the United States continued to operate its post–World War II dominance of shipping within the molds established by the earlier British regime. The huge excess capacity in the Great Lakes–sized boats that America built for World War II enhanced this continuity. It was not until the 1960s, when the Japanese set out to solve their joint problems of high import dependency and great distance to resources, that there occurred an increase in vessel size, complexity of port systems, and world tonnage equivalent to the one led by Britain at the turn of the century—only this time on a scale almost twenty times larger.

Hobsbawm (1968) argues that the conquest and administration of empire fed into the true industrial development of England, which occurred as its advantages in coal and iron fed its dominance of the new transport forms that emerged in the nineteenth century—rail and steamship. We have already argued that Britain's naval requirements stimulated a dynamic metallurgy that by 1800 gave Britain its enormous lead in metal-based industry. The British economy was only able to integrate these into competitive dominance, however, when transport technologies and the raw materials requirements of industry provided her with a relative advantage in access to the raw materials required for transport efficiency. Britain remained the most powerful shipping nation, and steamships motivated the rapid technical development of engine technologies—but the timing of the steamship revolution coincided with the end of British industrial development and the rise of other critical transport and production techniques. Shipbuilding stimulated a number of ancillary industries and skills in Britain, but it did not serve as dramatically as a template for new domestic industrial forms as it had in the less industrial Dutch economy. Transport strategies and infrastructures, on the other hand, did serve as a template for international organization, at the same time that the shipping industry retarded Britain's industrial decline.

The greatly increased scale and concentration of shipbuilding, together with the progressively diminishing share of raw materials transport in total value produced, means that transport's role as a generative sector is not likely to endure as long as it did in Holland and later in New England. Nonetheless, our analysis of the Japanese experience demonstrates that the continued close association of shipping, shipbuilding, and raw materials procurement with rapid industrial growth remains crucial in the creation of trade dominance and in the restructuring and spatial expansion of world markets.

The Japanese economy—first with U.S. state support but then increasingly in competition with the U.S. economy—expanded the scale and efficiency of maritime transport sufficiently to globalize world markets for even the bulkiest raw materials. Transport efficiencies based on vast increases in scale fundamentally shaped and supported the ways that the steel industry, the shipbuilding industry, and the planning and financial agencies of the national state, committed to working together to procure cheap and stable raw materials, generated Japan's internal and external economic power. The combination of exclusion from more proximate sources and technological and financial assistance to overcome the extra cost of distance for access to raw materials— both the fruit of U.S. geopolitical strategies—enhanced the globalizing consequences of Japan's industrial reconstruction.

The scale of Japan's globalization of raw materials transport is particularly striking. The following tables show how much the size of ships and the volumes and distances of raw materials transported increased during Japan's ascent to world dominance in both sectors and that the material intensification and spatial expansion that drive globalization still appear most dramatically in the transport of raw materials.

In other words, in 41 years, average tonnage almost tripled, the number of ships increased by a factor of 12, and total dry-bulk tonnage increased by a factor of 22. The long lifetime of boats meant that the average size increased much less dramatically than the maximum size: by the mid-1980s, Japanese yards were constructing ships of over 350,000 dwt. The *Berg Stahl,* for example,

Table 6.1. Number, Total Tonnage, and Average Size of World Dry-Bulk Carriers

Year	Number of Ships	Total Tonnage (dwt in millions)	Average Tonnage
1961	471	8.7	18,495
1970	1964	55.1	28,055
1980	4020	137.7	34,254
1990	4730	202.7	42,854
2000	5391	269.2	49,935
2002	5554	289.8	52,179

Source: Fearnleys World Bulk Fleet, Fearnleys World Bulk Trades, Fearnleys Review, and UNCTAD, various years.

Table 6.2. Transported Petroleum, Iron, and Coal in Thousands of Metric Tons

Year	Petroleum	Iron Ore	Coal
1960	366,000 (1962)	101,000	46,000
1970	995,000	247,000	101,000
1980	1,320,000	314,000	188,000
1990	1,190,000	347,000	342,000
2000	1,608,000	454,000	523,000

Source: Fearnleys World Bulk Fleet, Fearnleys World Bulk Trades, Fearnleys Review, and UNCTAD, various years.

jointly owned by CVRD and Japanese steel companies and specifically designed for the journeys that would connect Carajás to their steel mills, has a capacity of 385,000 dwt.

In terms of commodities, volumes of the most used raw materials increased between four and twelve times in forty years.

In terms of distances, for each of the same three raw materials, the ton-miles transported increased between five and nine times over the same period.

Like earlier ascendants to trade dominance, Japan's competitive success was based on achieving the cheapest landed cost of crucial raw materials. Unlike the Dutch and New Englanders had done, however, Japan did not achieve this by enhancing a naturally produced advantage. Rather, it relied on U.S. financial and diplomatic support to build the super-efficient ships and ports that supported its economies of scale in raw materials transport. Nor did Japan emulate the British and develop a powerful military apparatus to compensate for its lack of natural advantages. Instead, it remained protected by U.S. military might. Thus, financially, materially, geopolitically, and militarily, Japan remained dependent while it constructed a precarious, highly leveraged finan-

Table 6.3. Transported Petroleum, Iron, and Coal in Billions of Ton-Miles

Year	Petroleum	Iron Ore	Coal
1960	1650	34	264
1970	5597	1093	481
1980	8219	1651	957
1990	6261	1978	1849
2000	8180	2545	2509

Source: *Fearnleys World Bulk Fleet, Fearnleys World Bulk Trades, Fearnleys Review,* and UNCTAD, various years.

cial system subject to government intervention. This precarious dependency made Japan highly vulnerable when U.S. business started to see Japan as a dangerous competitor and when the U.S. government started to see it as a source of funds to solve its own fiscal deficit. As we noted earlier, this vulnerability exacerbated the prejudicial effects of Japan's raw materials access strategies on its peripheral suppliers. The fact that Japan's complex financial and industrial policies developed under the aegis of the United States ultimately brought economic crisis and stagnation to its own economy and deepened the impoverishment of its suppliers.

Anomalies of Dependent Trade Dominance

Japan's extraordinarily rapid rise, its precipitous decline, and the extreme negative impacts both had on the economic, social, and political integrity of the nations that supplied it with raw materials were in important ways all enhanced by the fact that the primary impulses that drove its economic boom and its decline were geopolitical rather than spatio-material. U.S. policies simultaneously impeded and enhanced Japan's ability to finance and supply three critical capital-intensive industries—shipbuilding, shipping, and raw materials procurement—at the scale necessary to be internationally competitive. Japan figured importantly in U.S. geopolitical strategies, and the Departments of Defense, State, and Commerce all saw Japan as the northernmost bulwark against the southward spread of communism into East Asia. They promoted it as a model demonstrating the benefits of capitalism to other Asian nations (Cummings 1984; Woo 1991). The U.S. proscription of commerce with the communist countries prohibited Japan from procuring raw materials from its prewar suppliers in North Korea, Manchuria, and China. The increased cost of raw materials clearly impeded Japan's ability to finance the reconstruction of its steel and shipbuilding industries. The daunting natural obstacle posed by Japan's lack of essential raw materials and her distance from adequate sources was thus geopolitically magnified by the same nation that created the demand for the expanded transformation of raw materials into marketable commodities.

The nation that created this combination of market opportunities and geopolitical constraints, though, also enabled the Japanese to develop and implement technologies at the scale required to resolve the inherent diseconomies of space it imposed on them. The United States also intervened to assist Japan in overcoming the spatial impediments that U.S. geopolitical strategies had exacerbated in the first place by providing financial, military, and diplomatic assistance in gaining access to large, high-quality mineral deposits in Australia and elsewhere and to technical and financial support in devising transport on the scale required to make such distant sources competitively viable. Even so, it was only possible for Japan to solve its raw materials problem by implementing far larger and more costly new technologies and economies of scale in transport and in steelmaking than would have been necessary if Japan had still had access to the Korean, Chinese, and Soviet raw materials markets. The economies of scale that U.S. geopolitical strategies imposed on Japan totally globalized the iron and coal markets, thus essentially globalizing world industry. They also made Japan world-dominant in those sectors where these massive economies of scale had been necessary.

Under the aegis of U.S. diplomatic, military, and financial support, the Japanese thus culminated historical processes of globalization by globalizing two of the lowest value-to-volume industrial inputs. The increases in the scale and scope of raw materials transport and procurement under three decades of unequal partnership between Japan and the United States were vastly greater than either could have achieved alone—indeed, greater than those achieved in any three-decade period of the far more equal interdependence of Britain and the United States. Their extreme speed—and evident precariousness—may well be due to the fact that their roots were geopolitical rather than spatio-material. To compensate for the absence of particular spatio-material advantages, the Japanese had to institute extremely large increases in economies of scale. To overcome the financial and organizational barriers the necessary technical scale imposed, they had to create more tightly coupled relations between state and capital than any nation had previously developed. The Japanese state devised extraordinarily close and effective ways to collaborate with national, financial, and industrial capital not just to overcome these obstacles, but also to use the technological, organizational, political, and financial innovations that the state-capital partnership developed to restructure radically world raw materials, product, and money markets in Japan's favor and against the interests of peripheral raw materials suppliers.

U.S. geopolitical campaigns to restrain China, North Korea, North Vietnam and insurgent movements in Laos and Cambodia as well as, of course, the Soviet Union engendered continued tactical support, military protection, and huge demand for military equipment and supplies. These effects of U.S. foreign policy made available additional finance for implementing technologies of scale in smelting and in downstream industry and easy, expansive markets in which to sell the increased volume of products they generated. These innovations, combined with the material and spatial conditions for which they were developed, catalyzed the globalization of investment and trade in even the bulkiest, lowest value-to-volume industrial inputs, in the process imposing extremely low rents and extremely high debt loads on extractive economies. In other words, the globalization that resulted from the geopolitical interdependence of Japan and the United States rested on an unsustainably precarious and overextended financial system in the core and on impossible levels of debt and excess extractive capacity in the periphery.

CHAPTER SEVEN

Conclusion

Few analysts attempt to link theories of national economic development, explanations of uneven economic development, and descriptions of the shifting and increasingly unequal structure of the world-system. Most that search for explanations of why some few economies manage to rise exceptionally rapidly emphasize unitary variables functioning in a unilinear fashion (see Goldstone, forthcoming). Explanations of why most economies fail to develop use similarly linear nomothetic approaches or remain ideographically focused on a single case. We have seen in the inquiry we are now concluding, though, that linear, nomothetic perspectives mystify uneven development. Rather, multiple processes and attributes—natural and social—intersect in locationally and temporally specific configurations to create the vigorous development and ascent of some national economies and the relative impoverishment, ecological and social, of others. We have also seen how the cumulatively sequential consequences of these processes progressively globalized the world economy and exacerbated the inequalities within it. Noticing that the competition to expand production, dominate trade, and increase consumption in order to generate profits and keep social order consistently catalyzed these historically reiterated processes of material intensification and spatial expansion, we wonder what form this competition will take and what its consequences will be as we reach the global limits of space and matter.

In this final chapter, we will review what we have learned from our inquiry and how we learned it. We will then consider what further analysis and what kinds of action we, as citizens and as scholars, can undertake to reduce the social and environmental disruption and destruction that broadening searches for raw materials have engendered. Finally, we will start the arduous task of trying to imagine a world-system in which nations use means other than material intensification and spatial expansion to solve their social, political, and economic problems.

Commonalities in and Transitions between Hegemons in an Expanding World-System

Each ascendant national economy transformed world markets for the most voluminously used raw material of the time—whether wood, iron, or steel—in ways that facilitated and cheapened their access to sources. Each in turn expanded the spaces across which they imported these raw materials—from river basin and coastal to lake-river-canal systems and from continental to oceanic and global ones. Each adopted and adapted new technologies and constructed new infrastructures for transporting and handling the bulkiest raw materials at greater scales and volumes than had previously existed. Each developed new financial and accounting instruments and new forms of firm or corporate organization. In each case, the state, firms, and financial institutions devised new, denser, more closely connected forms of collaboration and mutual monitoring and regulation.

Each thus achieved cheaper, larger, and more stable supplies of these raw materials than any competitor had managed. Each was able to convert this advantage in raw materials access, quality, and cost into competitive production for and competitive transport into international markets. Their competitiveness in product markets increased throughput, especially of the basic raw materials, and reduced the unit costs of the transport and processing infrastructure built to handle them. These savings, and the enhanced competitiveness in product markets that they enabled, supported further technological, financial, and organizational innovations that led to further economies of scale.

Economies of scale in the procurement, transport, and processing of the most voluminously used raw materials reduced input costs for a broad range of downstream manufacturers. Transport to and from, and the expanded commercial relations with, supplier nations extended and facilitated markets for these same manufactures. This created incentives for states and a broad coalition of national firms to collaborate in the huge sunk investments that the new technologies and expanded scale of infrastructure required. Because cheap and stable raw materials supplies also cheapened the construction of transport vessels and vehicles and of the rails, ports, and loaders needed to maintain the expanded flow of raw materials into the increasingly competitive heavy industries, each of these same economies in turn dominated the world production of transport vehicles. Successful state-sector-firm collaborations strengthened ties between and within all three and legitimated the growing authority they gave the state.

Successful national ascent to world economic dominance thus depended on a complex coordination of materio-spatial and social processes. Technology, as mediating between society and nature (Marx 1867; Harvey 1983; Hugill 1994), played a critical role in each ascent, but technological innovation, the investments it requires, and the mobilization of political and financial institutions around technology and finance are historically contingent within and dependent on multiple, far more regular material processes as these intersect with the evolving world-system at the historical moment of each ascent.

The cumulative sequence of these new technologies to cheapen the movement of greater volumes of matter across broader expanses of space drew on and exploited ever-broader topographical, hydrological, and geological features of the earth. The earliest cheap transport of bulky material depended on oar, sail, and raft to move down navigable rivers and on to relatively short coastal voyages. Coal-driven steamboat and rail transport later allowed the integration of lakes and ocean shorelines with much broader terrestrial hinterlands. In the last century, petroleum-powered supertankers and dry-bulk carriers made of high-tensile steel traverse global oceans while keeping the unit costs of the raw materials they transport sufficiently cheap to underprice smaller cargos brought across shorter distances (Ciccantell and Bunker 2002). The progressive expansion of commercially integrated space, first from the Mediterranean to the North Sea, then to the integration of the Atlantic and Indian Oceans, and next to the U.S. incorporation of the Great Lakes en route to the transcontinental linking of the Atlantic and Pacific Oceans, culminated in Japan's incorporation of global oceans as the medium of transport that globalized Japan's—and now China's—sources of iron, coal, and oil.

The nations that achieved the most rapid economic growth of their respective eras responded to the contradiction between economies of scale and the diseconomies of space by mobilizing and coordinating (1) technological and organizational innovation in domestic production and in transport, (2) construction and organization of their own and the world's built environment to cheapen the procurement and transport of the most voluminously used raw materials, and (3) reorganization of the political and financial relations between extractive suppliers and themselves. The rise to trade dominance in each case was technically, financially, and organizationally more demanding, and more rapid, than the one before it; involved sequentially cumulative increases in space and matter; and therefore required new, more closely coupled institutions of governance and collaboration between firms, banks, and the state, which in each iteration were more closely linked.

In each case of the most rapid economic ascent, the technological and organizational solutions to the contradiction between scale and space and the resulting problems of cheap and steady access to raw materials (a) enhanced the economic and political power of the ascending national economy while encumbering the developmental potential of its suppliers, (b) allowed the ascending nation to reshape patterns of trade in the most voluminously used raw materials to its own advantage and thus dominate world trade, and (c) expanded the flow of matter and integrated the processes of production over greater portions of the globe.

A Synthetic Summary of Historic Interactions between Extraction and Production

These repeated patterns constitute six centuries of cumulatively reiterated cycles, in each of which expanded economies of scale produced diseconomies of space that were resolved by even greater economies of scale. In order to implement these economies of scale, nations competing for trade dominance developed new and more powerful technologies, financial institutions, and state systems domestically. Abroad, they reorganized raw materials markets and transport systems in ways that complemented their domestic innovations and made them even more powerful.

Because social production can expand more rapidly than the natural production from which it draws matter and energy and because expanding production requires greater volumes and diversity of materials, each nation that ascended to trade dominance had to devise ways to obtain stable access to larger and more distant sources of more different kinds of matter. Each such nation extended the space within which it integrated and dominated extraction, production, exchange, finance, and international relations. New territories were incorporated according to the material and commercial needs, and for the political and economic benefit, of nations competing for trade dominance.

The technological, financial, and political systems such nations devised to expand existing economies of scale grew rapidly and accumulated power and wealth because they enabled the nations that developed them to draw on natural production from more different ecosystems. In contrast, each different local economy created to supply them with raw materials grew and accumulated power and wealth more slowly because each depended on natural production only in its own ecosystem and so tended to overharvest, degrade, deplete, and ultimately overshoot its material base.

Historically, these very different dynamics progressively increased the inequality between the productive, financial, and political powers of extractive and of productive economies; inequalities within extractive economies; and each extractive economy's impacts on its local ecosystem. Increased scale and the associated rising costs of investment led regional economies to specialize in the extraction of one or a few types of raw material. As we saw in the Amazon, concentration of labor and capital on a narrow range of natural products accelerates their depletion. Depletion sets off collateral destruction of species and landforms and waterways dependent on or affected by the natural products being extracted. The resulting reduction in the rates and variety of natural production limits both subsistence and commercial opportunities, finally making further extraction unprofitable. These internal dynamics are exacerbated by extractive economies' absolute dependence on external trade. This dependence increases their vulnerability to external market shocks and price drops. These processes concatenate the internal ecological destruction and the economic instability and discontinuity inherent in extraction.

We saw at Carajás that the concentration of capital and political power has now progressed to the point that extractive firms can negotiate with the state to lower rents and to diminish environmental and social regulation. International capital and politics have weakened the bargaining position of states by so reducing transport costs that each state must compete with all other possible sources. The extractive society thus suffers from the growth of two forms of inequality, one in its external relations with productive economies and the other in the growing internal gap between the small number of large firms that control large extractive projects and the rest of the increasingly immiserated society.

These multiple, mutually reinforcing, and strongly entwined cycles all emerge from and are driven by two contradictions. The first is rooted in the paradox that social production can expand much more rapidly than natural production, even as it remains absolutely dependent on it. The reiterated expansion of social production thus overshoots, degrades, depletes, and destroys the material basis that enables it to exist. This paradox creates a second contradiction. Each increased economy of scale, by raising the volume and number of different raw materials that social production requires, creates a diseconomy of space in trade and transport as competitive nations seek to loosen the limits that the first contradiction imposes on their economies. Each solution of either of these contradictions depends on increased scale, thus exacerbating the inequalities between the natural and social systems that create them. This means that each reiterated solution drives spatially larger and materially more intense cycles and that each such expansion and intensification accelerates the approach to global material and spatial limits.

In order to learn how these cycles globalized the world economy and what effects they had on local economies, we examined the ways that local configurations of matter and space in a single very large, ecologically very complex, extractive periphery—the Amazon—intersected with the progressively larger and more powerful economic and political powers, goals, social organizations, and technologies of globally trade-dominant nations. The ways matter and space shape economy were relatively easy to detect in Amazonian extraction. When we searched for the origins of the technologies that drove the demand for extraction, we found them too in the local configurations of matter and space that shaped not just the economy but also the political and civil culture of each nation that succeeded in ascending to trade dominance. We determined that in each case of successful ascent, these local materio-spatial configurations created particularly favorable access to the raw materials needed to make the most efficient transport vessels and infrastructure technologically possible at the time, combined with particularly favorable access to the richest markets for these same raw materials.

The nations that managed to parlay these particularly favorable conditions into a successful economic ascent did so by responding to the challenges and opportunities—socially and naturally produced—that their environments presented through collective efforts to implement new technologies, to institute new forms of social, economic, and financial organization, and to rebuild the environments of importing and exporting nations alike in ways that enhanced the economic advantages with which they started. Success of these efforts helped them to institutionalize their new organizational forms domestically and to impose them on their trading partners internationally.

Each of those that did succeed—the five nations we considered—radically reorganized the world's social hierarchy, its productive and transport technologies, its commercial relations, and its raw materials markets in ways that heightened that nation's comparative advantage in trade. This enabled each of them to dominate world trade for a time on a scale, and with a scope, speed, material intensity, and spatial extent of production, that surpassed those achieved by previous trade-dominant economies.

The technological achievements—the Dutch *fluyt* or Britain's innovations in using coal and iron to produce steam-powered engines, for example—and the expanded world markets that engendered each such reorganization eventually created opportunities for other nations with even richer resource bases and even better locations to establish commercial control over even broader spaces. These materio-spatial conditions enabled them to use new technologies to achieve more competitive economies of scale than the established trade-dominant nation could. These nations then repeated the same process on a grander scale. They developed even larger technologies and economies of scale and built the more powerful and capable financial and governance systems they required. Again, the new technologies and economies of scale first consolidated dominance of world trade for the nation that devised them but eventually overshot that nation's materio-spatial base, thus opening the way to economic ascent for societies with more favorable access to even greater resource bases over even greater spaces.

In each cycle, implementing new technologies to expand existing economies of scale required capital that exceeded the levels of accumulation possible under the economic scale of the previous cycle. In every case, those nations that responded to the challenges and opportunities of their environments in ways that enabled them to compete for trade dominance first created new, larger, and more powerful financial instruments and agencies capable of uniting a larger number of separate competitive capitals across larger space and larger populations than had been possible in the earlier cycles. We traced this progressive increase in financial scale and in the territorial scope of the capitals they could combine, from local banks based on personal relations and trust between wealthy families in Venice and Genoa, to banks and exchanges chartered and regulated by municipal authorities in Amsterdam but able to guarantee letters of credit that could be used in long-distance trade, to a national bank in Britain able to guarantee a national currency for use in international commerce, to stock markets in the United States able to underwrite and sell bonds and equity to capitalize state-registered corporations that grew first into multidivisional firms and then into transnational corporations, and finally to multilateral, or international, banks that could collect, coordinate, and deploy capital from multiple nations for investments on the scale of Carajás.

Each time expanded trade and transport required new, more powerful financial institutions, the state systems that controlled finance—first municipal, then national, and finally multinational—acquired expanded power, capacity and competence proportionate to the increased scale of the investments required and used to further the projects most critical to trade dominance. It was state initiative and assurance that overcame the antagonisms and distrust of the owners of individual capitals sufficiently to combine them in these single large investments of unprecedented scale. States guaranteed the contractual rights and enforced the contractual obligations of the enlarged corporate entities these combined capitals made possible. State regulation of the directors of the enlarged corporate entities aimed to prevent abuses of their huge financial power and to restrict their ability to restrain their competitors' trade by extra-economic means.

In all cases, the state learned to support and regulate new financial forms at the same time that it and capital were inventing them. This created complex state-sector-firm-finance relations of simultaneous interdependence and resistance. State regulation was essential to expanding financial institutions, but firms and financial institutions could increase short-term profits by dodging or defrauding state regulation. The state had to learn to regulate without restricting national capital's ability to compete in and for world markets, and capital could use the threat to withhold investment in order to gain more room to maneuver and manipulate finance for additional short-term profit. The state's relation to capital thus remains adversarial within a context of absolute mutual dependence and is therefore constantly contentious and shifting.

We can identify socially constructed attributes—cultural, ideological, or attitudinal, for instance—in each of the populations whose nations rose to trade dominance that could account plausibly for the commercial ambitions, technical imagination, work ethic, disposition to trust, and willingness to submit to authority that appear to have been necessary conditions for each nation's rise to trade dominance. There is no clear way of knowing, though, whether these attributes existed independently of the challenges and opportunities each nation's environment presented. Whatever their pre-ascent condition may have been, though, our examination of each nation's rise has made it abundantly clear that these socially constructed attributes evolved toward compatibility with productive and profitable responses to the favorable materio-spatial configurations—environment—from which each successful ascendant started. These configurations presented the challenges and opportunities about which each nation accumulated knowledge, skill, and power, and from which it learned the rewards of improving and using them for economic growth. They also, of course, set the materio-spatial context to which any successful technological innovation had to conform.

Put another way, our comparison of the ways national societies interact with environments in devising and improving technological, political, social, and financial means of production and trade suggests that those socially constructed attributes that serve to exploit more effectively and profitably the environment in which a population finds itself appear to be reinforced by the rewards resulting from their application to human action. The evidence that these socially constructed attributes change as a society responds to its particular natural environment leads us to imagine that their original formation was equally responsive to that environment. In other words, we conclude from our historical comparison that the social, political, and cultural patterns of societies that rise to trade dominance were formed in interaction with naturally occurring materio-spatial configurations that favored and facilitated the socially constructed technological, financial, transport, and military systems that drove their ascent.

In this sense, the histories of the different ascendant nations confirm the expectation we formed by tracing the sequence of extractive economies in the Amazon—that economic development in general and ascent to trade dominance in particular result from complex interactions between natural and social processes mediated and transformed by increasingly powerful technologies. The interaction between the vast spaces and huge resource bases in the United States, for example, catalyzed by the coincidence of expansion across the American continent with the development of iron- and coal-based technologies and economies of scale, generated particular patterns of relations between the state, capital, labor, and land and particular patterns of expectations and acceptance of certain public and private behaviors that enabled—and changed with—the construction and commercial use first of canals, then of railroads, and finally of highways (Meinig 1998; Parker 1991). Each stage in the more effective incorporation of space into the materially intensifying national economy significantly changed processes, relations, and volumes of production and exchange along with the political and financial systems that supported and regulated them and the technological systems that made them work.

The U.S. case dramatically exemplifies the ways that the challenges and opportunities of local configurations of matter and space shape the political and economic systems that devise technical and organizational responses to them. In similar ways, Dutch technological, political, and financial responses to the challenges and opportunities of their location on a low flat plain where large river systems fed into the conjuncture of the Baltic Sea, the North Sea, and the Atlantic Ocean shaped national political and economic systems in ways that facilitated the collaborative efforts of the state, capital, and labor, first to make the local environment more amenable to social production and profits and then to drive the national economy toward trade dominance (Israel 1995). Centuries later, MITI's success in coordinating Japan's response to the challenge of overcoming the huge distances to available raw materials sources created cross-sectoral enthusiasm for enormous national projects, business acceptance of significant state control and coordination, and public commitment to the very high levels of household saving (at very low interest rates) that made it possible for the state to target steel and shipbuilding for subsidized loans (Murphy 1996) and to subordinate competitive firms to a unified negotiating stance with foreign raw materials suppliers (Bunker and Ciccantell 1995a, b).

These examples clearly show that cases of successful development—and the rarer cases of successful ascent even more so—depend on felicitous, and thus relatively rare, combinations of specific sets of favorable materio-spatial conditions with compatible socioeconomic attributes sufficiently flexible to adjust to ongoing changes in the social and natural environments at particular junctures of time and space in the evolutionary trajectory of the world economy. The specific materio-spatial features and the specific socioeconomic attributes that might thus combine vary across times and places, and we have no reason to doubt that they could combine along various trajectories. In other words, there is no way we can know if a particular set of national responses was the best possible at that time, only that it was adequate to the day. We can be much surer, though, that they must concatenate in sequences and interactions that reinforce each other toward increased technical, financial, and political power and competence domestically and abroad in order to achieve economic development, and even more if they are to achieve trade dominance.

Within this frame, we agree that expanded finance and more powerful politics are essential to a nation's rise to trade dominance and fundamental to the increased inequalities that accompany the globalization of world raw materials markets. We have seen, though, that finance and politics do not develop independently of material process, as Arrighi (1994) claims, nor are they the root causes of globalization, as Harvey (1983) implies. Rather, finance and politics expand and grow stronger in interaction with and in proportion to (a) the material economies of scale in transport and production, (b) the task of subordinating other societies to guarantee cheap and stable access to their raw materials, and (c) the need to guarantee international safety of passage for trade in both raw materials and finished products.

Historically, finance and politics grow to meet the investment requirements of transport as it reaches across greater distances, for the trade of finished goods and of the raw materials to be transformed into finished goods. In this sense, trade does not follow the flag, because in more fundamental ways, the flag—and the banks—follow into the raw materials–providing zones where trade needs them to be. Politics expands to coordinate construction of the increasingly large and space-filling transport infrastructure needed for expanded trade and production, to support and regulate the increasingly powerful financial institutions that these projects generate, and to protect both the transport systems and the financial institutions that they generate from violations of property rights and contractual obligations. The national and international laws and institutions that define property and contract are either negotiated or simply assumed and then enforced by the trade-dominant nation in order to achieve the secure access to matter and to the space in which it is produced. As soon as we insist that our explanations of economic growth and development and of their regionally unequal impacts must integrate natural with social, material with financial, and ecological with economic processes, we see that these activities generate and depend on financial and political innovation and expansion, but they originally emerge from and continue to engage naturally occurring material processes.

The same dynamic processes that lift a few nations to trade dominance also create growing inequality between and within all of the world's nations. Just as matter and space provide the advantages that social actors maximize and exploit in successful development strategies, matter and space provide the conditions that prevent extractive economies from developing. Extractive and transport systems are built and operate according to the raw materials needs of ascendant nations. They also create the new markets that absorb the ascendant economy's extra production. The wages and rents paid for local labor and land create the extractive economy's demand and the ships returning empty from exporting raw materials provide the cheap transport for the increasingly abundant and increasingly cheap commodities the ascendant economy produces. These economic forces combine with resource depletion to retard capital accumulation in extraction and to provide it for industrial production.

The infrastructural, financial, and political conditions that make new markets, new fields of investment, or cheap labor in the periphery available to core capital are rooted in raw materials transport systems. Economies of scale cannot function without expanded scales of extraction and transport, and capital is unwilling to finance these systems in areas outside its political control. Trade-dominant nations therefore create formal regimes of administration and finance that govern relations with their peripheral suppliers—colonization, free trade, rights to foreign direct investment, autonomous joint ventures, structural adjustment programs, and organizations to monitor world trade and to guarantee continued cheap and stable access to adequate and expanding supplies of raw materials. When first Britain and then the United States encouraged independence movements that would open for commerce and investment resource-rich territories closed by earlier colonial regimes (Cardoso and Faletto 1969; Bunker and O'Hearn 1992), they intervened to encourage the formation of new states in forms that would facilitate the resource contracts they sought.

Peripheral states were not only created entirely divorced from the experience of growing with their own economies and societies, they were also born dependent and weak. Instead of evolving in ongoing interaction with expanding industrial production, as the liberal nineteenth- and twentieth-century states of Europe and North America did, imported forms of the contemporary state, modified to expedite resource extraction and export, were imposed on Asia, Latin America, and Africa. Lacking the deep historical experience of interaction with the social, political, and financial institutions of industrial capitalism, such states had inadequate competence, capacity, or autonomy to confront the rapacious appetites, ideological astuteness, and bargaining skills of the states and firms that sought access to resources in the territory these imposed states controlled. The present conflicts in oil-rich Middle Eastern states reflect both the peripheral instabilities of such impositions and the determination of core nations to stabilize them.

These political processes enhance the bargaining advantages that the ever-larger economies of scale and increased political power provide to capital as it searches for raw materials. Expanded scale and distance simultaneously increase the amounts and complexity of information needed to negotiate effectively in world markets and make it easier to conceal such information from peripheral suppliers. The relative inexperience of local state agents with the technical, financial, political, and legal complexities of large extractive projects further lessens their bargaining capacity and thus weakens their ability to palliate or demand adequate compensation for the social and environmental disruption these projects occasion. Depending on single large enterprises for tax revenues further limits the abilities of peripheral states to negotiate either resource rents or environmental protection, and it often enables firms to renegotiate their rent and tax obligations downward. CVRD was able to use the local state's dependency on its resource rents and taxes to bargain them down in return for completing fully the projects it used to win authorization in the first place but had then put on hold, alleging market change!

Carajás, as the latest and largest iteration of this process, illustrates the most extreme discrepancies between the technological and financial powers of extractive enterprises and the social and political powers of the societies that occupy the territories in which they operate. This discrepancy reflects the even greater—and rapidly growing—inequality between the technological and financial powers of the industrial economies that compete for trade dominance and those of the extractive economies whose politics and policies core firms and states manipulate in order to lower their raw materials costs. This inequality means, in turn, that as world capitalism progresses historically by reducing the unit costs of production and transport, there is less and less chance that control over raw materials can be transformed into sustained competitive production and trade. Capital barriers to local production go up as economies of scale increase, and the tariff barriers of distance that once protected local industry come down as transport costs are reduced.

The financial and political processes many authors see as primary forces in structuring the world economy are rooted in capital's need for raw materials. Finance and politics are quintessentially social constructs, and they work powerfully to advance the purposes of the human groups that control them. Neither they nor the other social activities and attributes involved in economic development can function independently of matter and space, however. We cannot see this, just as we cannot see the increasing toll it imposes on social and environmental systems or the ways it increases inequality within and between societies—unless we find ways to engage "socio-logic" with "eco-logic."

There is no single set of attributes, actions, or processes that lead in anything like a linear progression to trade dominance or globalization, nor for that matter to decline from trade dominance. Rather, by comparing the cases of successful ascent and by accounting for similarities and differences in each case as the global economy intensifies and grows over time, we have abstracted a series of necessary conditions—material, spatial, social, economic, and financial—that may be met in various ways but always within the parameters of material process and physical law and must concatenate in mutually reinforcing sequences and cycles for strategies of national ascent to succeed.

It seems likely that the world-system is reaching the global limits of new locations to search for larger deposits of high-grade material. Global transport systems now involve more states in contests for access to raw materials and thus increase the likelihood and the intensity of violent struggle internationally —and violent repression domestically—to control them. The sequentially cumulative increase of technological capacity and scale at cheaper unit costs that expanded and accelerated world transformation of natural products into social products has tremendously increased the power and cheapened unit costs of weapons, further enhancing the probability and aggravating the environmental and human costs of international struggle and domestic repression. As these solutions reach their natural limits, capital's solutions to the contradiction of scale and space may engender even greater aggression and violence in addition to increased inequality and environmental destruction.

Hegemonic ideologies of the industrial core that equate comparative advantage in natural resources with comparative advantage in developing transformative industries are used to foment and then manipulate the hopes and expectations of resource-rich peripheries aspiring to economic growth and social progress. Fernando Coronil (1997) brilliantly documents the ways that peripheral politicians internalize and propogate ideologies of occidentalism and modernity that legitimate the access of transnational corporations to local natural resources. Coronil analyzes the perverse consequences that diffusion of this false consciousness has on the state's rent policies and development programs—and thence on civil society, the military, various social classes, the press, and political parties. Bunker (1989a) shows how mainstream theories of development became hegemonic ideologies that the national government in Brazil and foreign investors in the Amazon used to silence local objections to the Carajás mine and the Albras aluminum project.

Intellectuals, politicians, and capitalists in extractive regions clearly understand that industry depends on—and must adapt to—the use values in raw materials. They are prone to extrapolate their correct understanding that all value, profit, and wealth, though ultimately realized or produced by labor, originate in or on the ground to an expectation that whoever controls the ground that contains this "natural" value can obtain considerable profit and wealth from nations that need to import the material that lends substance to their labor. Ricardo's and Marx's theories of ground rent are based on exactly this assumption; both show rents proportionate to the contribution that differential rates of natural production, or "fertility," make to social productivity and to profit.

Hegemonic simplification and manipulation of these assumptions foster illusions that rents will provide the capital—and nature the inputs—needed to develop diversified local industry in the extractive region. These ideas were valid in the centuries past when they were formulated, but they became illusions as increased transport efficiencies forced resource-rich nations to compete for markets by lowering their rents. Theoretical and ideological distortions and manipulations of contemporary reality, though, are still deployed to convince local states and firms to assume capital risk for the extractive economies that will export their resources to make industrial economies elsewhere more profitable (Bunker 1994a). When the resulting excess capacity in extractive infrastructure and supply depresses world prices below operating costs, these extractive firms and states have to continue operating their mines, ports, and railroads at full capacity in order to generate the revenue flow required to service their debts. This keeps world supplies excessive, prices correspondingly low, extractive economies poor, and extractive states weak (Bunker 2000).

Our final analytical task in this book, then, is to consider the ways that core states and firms create and manipulate these illusions about the local benefits of extraction, so that we can start to imagine ways to initiate a sharing of information, ideas, and attitudes that might enable citizens and scholars in the productive core and in the extractive periphery to coordinate effective and knowledgeable campaigns to lessen the damage that globalized raw materials extraction and transport do to society and to the environment.

Unmasking the Illusion that Extraction Leads to Economic Development

Negotiations between the local state and giant mineral companies over the size, operation, property rights, and distribution of costs and benefits for large extractive projects proposed in virgin territory are uneven contests between unequal adversaries. The widely held notions that natural resources are a free gift of nature and that the nation in whose territory they occur is blessed with wealth in the ground make it relatively easy for the agents of business to promote the idea that extraction leads to development. Peripheral states are thus persuaded to authorize, promote, and in many cases share the costs of huge new extractive projects. Just as the growth in scale, distance, and technical complexity facilitates the mystification of the material and spatial relations that produce much of what they consume for citizens and scholars in the core, so too do they facilitate the mystification of the local costs and benefits of extracting matter from the space on which private and public agents in the periphery depend.

Thus, as distance, scale, and technical complexity increase, the contractual terms for new mines reflect the growing inequality in the bargaining capacity of huge mineral corporations in the core and agents of peripheral states. Negotiation of mineral contracts pits knowledgeable, experienced agents who have access to complex sets of specialized information closely held by the corporations they represent against agents of a state with little or no experience and with access only to general and often out-of-date information about the technologies and markets for the natural resource in question.

Whatever specific knowledge peripheral negotiators do have of mineral economies is generally drawn from abstract, widely accepted theories that lose whatever validity they might have had as time passes and transport efficiency increases. Less than a century ago, recent experiences of the United States, Canada, Australia, and South Africa suggested that raw materials sales provided capital for industrial development in the regions from which they were extracted (see Katzman 1987 for an encyclopedic review and invocation of these vent-for-surplus theories). Subsequent studies showed that, even then, this only occurred where extraction catalyzed commercial exploitation of other natural features, such as fertile lands, extensive forests, and access to navigable waters (Perloff and Wingo 1961; Caves and Holton 1959). Mainstream development theories, though, especially those espoused by the IMF, the World Bank, and the foreign representations of core economy interests such as U.S. and Japanese Export-Import Banks, USAID, JICA, and so on, ignore these subtle geophysical distinctions. They continue to draw on and to promulgate simplistic vent-for-surplus assurances that raw materials extraction and export will provide the capital and technology required for industrial development.

The fact that all of the national economies that rose to trade dominance before 1950 did so by incorporating high-quality raw materials cheaply available in or near their own territories into technologies that provided the most competitive economies of scale at that time, combined with the fact that the economies of scale in transport now function simultaneously to raise debt levels, lower raw materials prices, and retard local industry in the extractive region, raises serious doubts about the prospects of industrial development for economies dominated by the large-scale mineral projects that contemporary raw materials markets demand. We would propose instead that the most important thing that states and societies of extractive economies can do to improve their well-being and preserve their environments is to form international cartels around each of the raw materials they export in order to end the ongoing reduction in their rents and to break the emerging pattern of payment by the periphery for the transport infrastructure that will benefit only the industrial economies that import these raw materials. Such material-specific cartels could ultimately join in cross-commodity global systems of governance that would regulate the contract terms and environmental and social effects of extractive projects in order to distribute more equitably the complex mix of benefits produced by the interaction of matter and labor.

To present such a proposal successfully, we need to do analysis and make arguments persuasive enough to offset the blandishments of wealthy, powerful, resource-dependent states and firms in the core, which use conventional development theories that promise industrial development to persuade individual extractive economies to compete for market share by lowering rents and assuming debt to pay for transport infrastructure. We would also have to convince the nations of extractive economies to band together to contest the free trade arguments that core nations and multilateral banks use to justify their opposition to cartels organized to protect natural resource rents and the environments from which these resources are drawn.

In doing this, we will need to deal with the problem that these theories were all developed in and are all based on data from successful industrial econo- mies. In most peripheral nations, the most influential economists are those trained in the core in theories that emerge from—and ultimately benefit—the core. We saw in the Amazon that CVRD, along with those ministries of the national state that supported its projects in Carajás, used these economists to argue against local state attempts to restrict CVRD's power in the region. Overcoming these kinds of vested power will require careful analysis of a broad range of data within the theoretical frame we have developed in this study.

In the following section, we consider what changes in the social sciences and in public awareness might support a fuller appreciation of the role of natural production in human industrial economies, changes that in turn would enable global collaboration to achieve global improvements in the terms of contracts and in the rates of extraction they regulate.

How the History of Capitalism Explains the Poverty of Extraction

The proximity of nineteenth-century industry to nineteenth-century ex- traction created the illusion that the wealth in the ground was the source of regional and national development, but correlating development and extrac- tion is spurious even for that time and a dangerous illusion for the twenty-first century. Cheap, abundant, high-quality raw materials underlay the huge surge in productivity, but it was labor shortages, labor militancy, and higher wages that drove development in the United States. The correlation of extraction and development is spurious (Wright 1990). It is quite clear that the flow of Bra- zilian raw materials into Japanese industrial products is not developing the Amazon, and the only changes in the relation of social to natural production over the intervening century and a half were in the scale of extraction, the scale of transport, and the volumes and distances involved.

Capitalist institutions do not spring full-blown to match contemporary scales of extraction; rather, they evolve in interaction with social and tech- nological innovations, compatible with the favorable materials locally avail- able, that produce expanding economies of scale in production. This logic implies that the social and environmental advantage remains with those nations that combine favorable access to raw materials with developing institutions of competitive capitalism. As economies of scale drive raw materials extraction toward global limits, ecologically fragile areas of the tropics and the arctic, where capitalist institutions and states are least developed, are required to bear their environmental, as well as their social and economic, costs. In these processes as well, we see how distance combines with depletion in generating sequentially cumulative inequality between extraction and production.

We have seen how natural and social dynamics engendered increasing inequities and losses. What can we, as scholars and citizens, do to reduce them? The first and the most obvious step would be to create cooperative networks of those concerned in the core and in the periphery. These networks can gather data about and coordinate analysis of ongoing and planned projects of core states and firms—ranging from military campaigns to subversive actions to ideological manipulations to various forms of negotiation and investment—to secure favorable access to raw materials. This information and analysis can then be shared with citizens, state agents and negotiators, and social movements in the countries whose raw materials are being sought, first to correct the mystification and alienation that some participants in these negotiations may use to gain cheap access to natural resources and then to strengthen the bargaining positions and demands of participants seeking to protect the human rights, social well-being, and environmental integrity of the various life systems that the proposed extractive enterprises would affect.

To be effective, such analysis must examine the diverse local configurations and multiple intersecting systems that comprise the global system with as much attention to natural and social detail as we have given to our historical comparisons. In other words, we must take into account the full complexity of the processes that create—and those that temporarily solve but then recreate on a larger scale—the contradiction between natural and social production. This will require collection and sharing of scientific knowledge about specific natural production in many different places, together with collection and sharing of technical and financial information about the large extractive projects being planned and negotiated. Such an endeavor will need huge imagination and huge effort coordinated across multiple specialized skills and disciplines and across multiple national boundaries sustained for the length of the actions and outcomes being analyzed.

This is a daunting task, but the work required could be intrinsically rewarding. The diversity and complexity of local natural systems and of the societies that adapt to them can stimulate joy, awe, and love in the attentive, persistent observer. The natural world is a beautiful and exciting place to know, and different societies develop wonderfully ingenious ways of knowing it. Different disciplines of specialized knowledge can articulate and communicate with general knowledge indigenous to specific ecosystems in ways that enrich both kinds of knowledge and delight the different kinds of knowers. In the next section, we will consider what kinds of communication could develop across these multiple boundaries of nation, ecosystem, and special forms of knowledge.

What Is to Be Done?

Marx wrote once that humans make their own history, but not as they intend. Tracing as we have the history of expanded financial, political, and social inequalities between extractive and productive societies and of the expanding social and environmental disruptions of large-scale extraction from its origins in the Mediterranean to its contemporary apex in the Amazon leads us to reject Marx's phrase as too comfortable and too easy, both analytically and ethically. Rather, we, as citizens and scholars, must strive to assure that humans consider the history their actions will make while deciding on those actions. Using history to show how contemporary extractive projects both perpetuate and shape the ongoing contradictions between economies of scale and diseconomies of space and realizing how the reiterated solutions to these contradictions create enduring environmental, political, and social damage, we are confronted with two different tasks. First, what can we, as scholars and as citizens, do to ameliorate the inequities and distortions visited upon those places that have supplied the raw materials that core capitalist economies have transformed into wealth, power, and privilege? Second, what influence can we have over the negotiation and implementation of present and future projects to supply the raw materials that core capital will continue to demand?

As authors and as readers of this book, we have already taken the first steps toward embarking on these tasks. States and firms, in both extractive peripheries and in the industrial core, promulgate mystifying ideologies that obscure the local costs and damages of extraction and promise that extraction will sustain economic growth and social development. These mystifying ideologies have become more effective as advancing technology and division of labor attendant on the progressive spatial separation of extraction from production have alienated most of us from basic material processes. To the extent that we insist on knowing and promulgating information about these processes, we can resist this mystification. In other words, we can use adequate and accurate historical accounts to counter the partial and distorted historical accounts that hegemonic discourses promulgate.

We must promote sustained efforts to encourage awareness of how varied material processes in socially and topographically differentiated local spaces constitute the global in order to demystify the discourses that justify the subordination and exploitation by core productive economies of environments that contain critical raw materials and of the populations that live in those environments.

A first line of opposition to the increasing inequality between extractive and productive economies would be to unveil the benefits claimed for modern extraction by making bargaining processes more transparent and more broadly public. Fortunately, the information technology that was devised to coordinate global military and financial programs has become cheap enough, and widely enough distributed, that it is quite possible to gain access to and divulge the conditions and the terms of contracts being negotiated. We can also analyze the costs and benefits being claimed for them by comparing any proposal with existing projects.

The recent transnational campaigns to protect particular environments, the achievements of transnational labor movements, and the more recent efforts to oppose multilateral trade agreements by a union of anti-sweatshop groups with environmental and human rights organizations have all been supported by the international operation of information technology that is available to a small but growing proportion of the world's population. Organizations such as the Transnational Institute in Holland have successfully organized international gatherings about participatory democracy and global human rights. Municipal boycotts of Shell Oil in protest of the environmental and civil rights damages it has occasioned in Nigeria or student protests against Freeport McMoran for assaults on civilian populations at its Grasberg mine in Indonesia testify to the willingness of citizens of the core to mobilize on behalf of citizens of extractive peripheries. There have been parallel consumer actions fighting for fair trade for peasant coffee growers, sustainable harvests of mahogany, or the capture of tuna without compromising associated species of fish and aquatic mammals.

Obviously, it is easier to capture people's imagination and sympathy about attractive species of plants and animals whose origins and sources can still be directly identified even after they appear as commodities. It is simpler to respond to the single kind of raw material from which your coffee is produced than to the myriad kinds of raw material that go into your Toyota. It will require huge coordinated and sustained effort to conduct global vigilance over the conditions of extraction of the full range of raw materials on which our extraordinarily high standard of living depends. The means exist, though, for informed collaboration between citizens and scholars of the core and of the periphery to share and make publicly available, and then to use to influence state decisions, information relevant to specific negotiations for new mines. We can also use them to instigate critical discussion of the morality and equity of different core state policies and strategies for cheap and stable access to militarily or industrially critical raw materials.

In order to gather and share such information effectively, though, we must demystify for ourselves, both as citizens and as scholars, the ways that society depends on nature and the ways technology is subordinate to naturally produced matter and space. This requires that we reorient our approaches to the social sciences—and then to social action—sufficiently to allow us to incorporate the full and glorious detail of localities into a systemic understanding of the global. We turn to that task in the next section.

Global Consequences of Successful Strategies to Dominate Trade

As we followed the history of the rise of successive nations to world trade dominance, the linkages among (1) the expanding economies of scale in extraction, transport, and production, (2) the growing capital barriers to entry as scale of investment expanded, (3) the relative loss of state autonomy and capacity as capital scale increased, (4) the increasing inequality between extractive peripheries and the industrial core, (5) the complicity of peripheral actors in core nation strategies to cheapen raw materials rents and devolve the costs of extraction and transport infrastructure on the state and civil society in the extractive economies, and (6) globalization become dramatically stronger and clearer.

We saw that the diseconomies of space that engender the sequentially cumulative expansions of technology, transport systems, financial institutions, and the powers of the national state are rooted in the unequal growth rates of natural and social production. This inequality leads eventually to the degradation and impoverishment of the ecosystems that provide the matter and energy that social production transforms. It also leads to the impoverishment of the extractive economies that mediate between natural and social production, because individual extractive systems are bound to the natural production of their particular ecosystems.

In order to see and analyze clearly how social production's absolute dependence on and growing ability to appropriate and exploit natural production leads both to globalization and to growing inequality between extractive and productive economies, we had to overcome modern social science's reluctance to problematize and to theorize matter and space. We found useful guidance in the writings of earlier social scientists more attuned to material processes of production. Applying some of their insights to today's more globalized and materially more complex economies allowed us to penetrate the ways that modern industry and the modern industrial state mystify nature and alienate us from it, first as consumers but also as citizens and as scholars. We could then devise an analytical frame that integrated the very different levels of analysis and generality required to examine the intersections of categorically opposite phenomena—material and financial, natural and social, physical and ideological, trans-temporal and conjunctural, contingent and universal—that are usually the domain of quite distinct intellectual disciplines using quite different theoretical and analytical approaches.

The analysis this frame enabled us to do showed the ways that the natural and social processes that drive the reiterated solutions to the recurring contradiction between economies of scale and diseconomies of space interact at the intersection of the self-accelerating growth of productivity in industrial economies and the self-reinforcing loss of productivity in the natural systems that extractive economies exploit. It also showed us that the growth of productivity in industry depends absolutely on the same processes that lead to the loss of product and of rents in extraction. The growth of this interactive and interdependent inequality between extraction and production fostered increased social disruption and environmental degradation in the resource-exporting periphery.

This analysis, though, makes clear a second and steadily growing deficiency of the modern social sciences, that is, their lack of attention to the ethical and moral dilemmas—environmental and social—that use of expanded volumes and types of matter traded and transported from multiple ecosystems to devise new products and more efficient technologies of transport and production inevitably cause to surface. Problematizing the ways that technology mediates between society and economy, on the one hand, and matter and space, on the other, brings this dilemma into sharp relief. Each instance of technological innovation that we considered in this account was an example of human ingenuity about, sensitivity to, and discovery of previously hidden dynamics of nature. Human curiosity and imagination about the natural processes that shape their environments were critical first to human subsistence and survival and then to steady improvements in welfare and living standards. We believe that human delight and awe in natural phenomena and human compassion for other humans were both fundamental to social evolution and to the technical advances that solved the Malthusian curse and broadened the possibilities for living longer, more comfortable, and more interesting lives—in at least some of the world's societies. In this sense, we must confront the paradox that the technical discoveries and social innovations we have described as exacerbating social and environmental disruptions and inequalities on an ever-broader scale are rooted in some of the noblest and most generous of human impulses.

The dilemma is that the technologies that emerge from these impulses make possible vastly increased power over the natural environment and over the human groups that draw their subsistence from it. As the scale of these technologies and the capital barriers to their invention and implementation increase, they become increasingly susceptible to capture and monopoly control by individuals and groups motivated by greed and desire for power. Thus, creations stemming from the noblest of human impulses increasingly serve the basest of human impulses.

The modern social sciences have avoided confronting this dilemma, despite ample precedent and example in the work in which they originated. The ethical and moral questions that the huge social and political changes and disruptions engendered by the accelerated rates of technological and organizational innovation and the expanded economies of scale and diversification that iron and coal made possible in the early nineteenth century prompted the seminal works of Marx, Weber, Durkheim, and numerous other still-influential analysts. Their theories, in turn, were rooted in earlier examinations—by authors ranging from Locke and Quesnay to Smith and Ricardo—of the ethical issues raised by the increased social power that long-distance trade made possible.

As scientific inquiry responded to the growing complexity and opportunity that the material revolution of the nineteenth century made possible by becoming more specialized and as technical and statistical procedures appropriate to specific kinds of inquiry increasingly bounded the goals and methods of scientific research, these ethical questions have remained outside the increasingly high walls that the social sciences have constructed between empirical, objective analysis and the ethics and principles that constrain human greed and its drive to power. In other words, as material technologies and their ambivalent capacity to enhance simultaneously human welfare and social oppression have become more powerful and as the increased scale of production and transport has increased the distance between extractive peripheries and core industries, research technologies have become more specialized and less apt to engage broader ethical issues of social inequality, political subordination, and environmental destruction.

The ethical dilemmas of the relations between industrial core and the extractive peripheries on which it depends are further complicated by the fact that the latest cycle of the economies of scale devised to offset the diseconomies of space may have come close to exhausting the still unexploited deposits of industrially critical raw materials large enough to sustain the next round of economies of scale. Ton-miles transported and the total volume of base materials consumed continue to grow more rapidly than world GDP. These data certainly raise serious doubts about the complacent claim that the world economy is somehow dematerializing and therefore treating the environment more gently (Bunker 1996). It seems more likely that economies of scale and the expanded volumes of raw material they require will continue to accelerate as the availability of large new sources rapidly diminishes. At the same time, Japan's current economic crisis serves as a warning that the financial and political innovations needed to confront the vastly increased investments required for the next cycle of economies of scale may be so complex and require such hypercoherent institutional relationships as to leave any nation-state that attempts them vulnerable to similar crises.

That China, with its contentious international relations and major unresolved domestic conflicts between a centralized and corrupt national state and a huge, culturally fragmented civil society, has since 2000 been the world's major importer of iron and declares plans for major infrastructural development in railroads and deep-water ports over the next decades (Bunker and Ciccantell, forthcoming) does not bode well for coherent global solutions to what may be the global culmination of the reiterated and self-expanding solutions to the contradiction between scale and space. That the United States, with its recent and increasingly visible expressions of contempt for multilateral institutions of international diplomacy and reconciliation and for agreements covering human rights, nuclear control, and environmental integrity and its increasingly blatant use of unilateral military force and economic sanctions to secure control over oil in the Middle East and elsewhere, remains unquestionably the most powerful nation in the world bodes no better.

In other words, the environmental and social problems caused by the growing disparity between the growth rates of natural and of social production and the reiterated contradiction between scale and space to which it leads may soon have accumulated to the global boundaries of natural production even as the political cultures developed in the most powerful nations that most powerfully drive the accumulating problems appear progressively less open to the kinds of international collaboration that might bring about alternative solutions or at least amelioration.

Such an expansive set of problems leaves us, as citizens and as scholars, facing overwhelmingly powerful forces with only intellectual power and moral suasion, but if we can combine sensitivity and imagination about the environment with openness to information about the social and cultural organization of other peoples and combine our efforts with theirs to persuade the multiple states of core and periphery that continued depredation of the natural environment is an unacceptable consequence of the historically reiterated solution of using economies of scale to overcome the diseconomies of space, we can start to globalize civil society across multiple diverse nations as effectively as capital has globalized trade across multiple diverse ecosystems. One obvious and very important first consequence of such a movement would be to increase enormously the prices of raw materials, thus reducing their consumption in the core, reducing the inequality between core and periphery, and reducing our destruction of the natural environment.

References

AISA. 1902. *Statistics of the American and Foreign Iron Trades for 1901.* Philadelphia: American Iron and Steel Association.

AISI. Various years. *Annual Statistical Report.* Philadelphia: American Iron and Steel Institute.

Albion, R. 1926. *Forests and Sea Power: The Timber Problem of the Royal Navy, 1652–1862.* Cambridge: Harvard University Press.

Arrighi, Giovanni. 1994. *The Long Twentieth Century.* London: Verso.

Ashton, T. S. 1924. *Iron and Steel in the Industrial Revolution.* Manchester: Manchester University Press.

Barbour, Violet. 1930. "Dutch and English Merchant Shipping in the Seventeenth Century." *Economic History Review* 2, no. 2: 261–90.

———. 1950. *Capitalism in Amsterdam in the Seventeenth Century.* Ann Arbor: University of Michigan Press.

Barham, Bradford, Stephen G. Bunker, and Denis O'Hearn, eds. 1994. *States, Firms, and Raw Materials: The World Economy and Ecology of Aluminum.* Madison: University of Wisconsin Press.

Barker, Preston Wallace. 1938. *Rubber Statistics, 1900–1937: Production, Absorption, Stocks, and Prices.* Washington, DC: GPO.

Beard, Charles. 1930. *The Rise of American Civilization.* New York: Macmillan.

Becker, David G. 1983. *The New Bourgeoisie and the Limits of Dependency: Mining, Class, and Power in "Revolutionary" Peru.* Princeton: Princeton University Press.

Berglund, Abraham. 1931. *Ocean Transportation.* New York: Longman.

———. 1934. "Shipping." *Encyclopaedia of the Social Sciences,* 14:30–43. New York: Macmillan.

Bidwell, Percy. 1958. *Raw Materials: A Study of American Policy,* New York: Harper Bros.

BISF. 1951. *Statistics of the Iron and Steel Industries of Overseas Countries for 1950.* London: British Iron and Steel Federation.

Bomsel, Olivier, et al. 1990. *Mining and Metallurgy Investment in the Third World: The End of Large Projects.* Paris: OECD.

Bosson, R., and B. Varon. 1977. *The Mining Industry in the Third World.* Washington, DC: World Bank.

Boxer, C. R. 1965. *The Dutch Seaborne Empire: 1600–1800.* London: Hutchinson.

Braudel, Fernand. 1984. *The Perspective of the World.* New York: Harper and Row.

Brenner, Robert. 1993. *Merchants and Revolution: Commercial Change, Political Conflict, and London's Overseas Traders.* Cambridge: Cambridge University Press.

Brockway, L. 1979. *Science and Colonial Expansion: The Role of the British Royal Botanic Gardens.* New York: Academic Press.

Bunker, Stephen G. 1985. *Underdeveloping the Amazon: Extraction, Unequal Exchange, and the Failure of the Modern State.* Urbana: University of Illinois Press.

———. 1986. "On Values in Modes and Models: Reply to Volk." *American Journal of Sociology* 91, no. 6: 1437–44.

———. 1989a. "The Eternal Conquest." *NACLA Report on the Americas* 23, no. 1: 27–35.

———. 1989b. "Staples, Links, and Poles in the Construction of Regional Development Theories." *Sociological Forum* 4, no. 4: 589–609.

———. 1992. "Natural Resource Extraction and Power Differentials in the World Economy." In *Understanding Economic Process,* ed. Sutti Ortiz and Susan Lees, 61–84. Washington, DC: University Presses of America.

———. 1994a. "Flimsy Joint Ventures in Fragile Environments." In *States, Firms, and Raw Materials: The World Economy and Ecology of Aluminum,* ed. Bradford Barham, Stephen G. Bunker, and Denis O'Hearn, 261–96.

———. 1994b. "The Political Economy and Ecology of Raw Material Extraction and Trade." In *Industrial Ecology and Global Change,* ed. Robert Socolow, C. Andrews, F. Berkhout, and V. Thomas, 437–50. Cambridge: Cambridge University Press.

———. 1996. "Raw Materials and the Global Economy: Oversights and Distortions in Industrial Ecology." *Society and Natural Resources* 9: 419–29.

———. 2000. "Notas sobre a renda do solo e a tributação." *Papers do NAEA* 159.

Bunker, Stephen G., and Paul Ciccantell. 1994. "The Evolution of the World Aluminum Industry." In *States, Firms, and Raw Materials: The World Economy and Ecology of Aluminum,* ed. Bradford Barham, Stephen G. Bunker, and Denis O'Hearn, 39–62.

———. 1995a. "Restructuring Markets, Reorganizing Nature: An Examination of Japanese Strategies for Access to Raw Materials." *Journal of World-Systems Research* 1.

———. 1995b. "Restructuring Space, Time, and Competitive Advantage: Japan and Raw Materials Transport after World War II." In *A New World Order? Global Transformation in the Late 20th Century,* ed. David Smith and Josef Borocz. Westport, CT: Greenwood Press.

———. 2003a. "Generative Sectors and the New Historical Materialism: Economic Ascent and the Cumulatively Sequential Restructuring of the World Economy." *Studies in Comparative International Development* 37, no. 4: 3–30.

———. 2003b. "Creating Hegemony via Raw Materials Access: Strategies in Holland and Japan." *Review* 26, no. 4: 339–80.

———. Forthcoming. *Restructuring Markets and Reorganizing Nature: The Political Economy and Ecology of Japan's Global Search for Raw Materials.*

Bunker, Stephen G., and Denis O'Hearn. 1992. "Strategies of Economic Ascendants for Access to Raw Materials: A Comparison of the U.S. and Japan." In *Pacific Asia and the Future of the World-System,* ed. Ravi Arvind Palat, 83–102. Westport, CT: Greenwood Press.

Cafruny, A. 1987. *Ruling the Waves: The Political Economy of International Shipping.* Berkeley: University of California Press.

Cardoso, Fernando, and Enzo Faletto. 1969. *Dependencia y desarrollo en América Latina.* Mexico City: Siglo Veintiuno.

Caves, Richard E., and Richard H. Holton. 1959. *The Canadian Economy: Prospect and Retrospect.* Cambridge: Harvard University Press.

Chandler, Alfred D. 1962. *Strategy and Structure: Chapters in the History of Industrial Enterprise.* Cambridge: MIT Press.

———, ed. 1965. *The Railroads: The Nation's First Big Business.* New York: Harcourt Brace and World.

———. 1977. *The Visible Hand: The Managerial Revolution in American Business.* Cambridge: Harvard University Press.

———. 1990. *Scale and Scope: The Dynamics of Industrial Capitalism.* Cambridge: Harvard University Press.

Chase-Dunn, Christopher, Yukio Kawano, and Benjamin Brewer. 2000. "Trade Globalization since 1795: Waves of Integration in the World-System." *American Sociological Review* 65, no. 1: 77–95.

Chew, Sing C. 1992. *Logs for Capital: The Timber Industry and Capitalist Enterprise in the Nineteenth Century.* Westport, CT: Greenwood Press.

Chida, T., and P. Davies. 1990. *The Japanese Shipping and Shipbuilding Industries.* London: Athlone Press.

Ciccantell, Paul, and Stephen G. Bunker. 1997. *Space and Transport in the World System.* Westport, CT: Greenwood Press.

———. 1999. "Economic Ascent and the Global Environment: World-Systems Theory and the New Historical Materialism." In *Ecology and World-Systems Theory,* ed. Walter Goldfrank et al. Westport, CT: Greenwood Press.

———. 2002. "International Inequality in the Age of Globalization: Japanese Economic Ascent and the Restructuring of the Capitalist World Economy." *Journal of World-Systems Research* 8, no. 1: 62–98.

Cipolla, Carlo M. 1965. *Guns, Sails, and Empires: Technological Innovation and the Early Phases of European Expansion, 1400–1700.* New York: Minerva Press.

Coronil, Fernando. 1997. *The Magical State: Nature, Money, and Modernity in Venezuela.* Chicago: University of Chicago Press.

Couper, Alistair. 1972. *Geography of Sea Transport.* London: Hutchinson University Library.

———, ed. 1983. *The Times Atlas of the Oceans.* New York: Van Nostrand Reinhold.

Cummings, Bruce. 1984. "The Origins and Development of the Northeast Asian Political Economy: Industrial Sectors, Product Cycles, and Political Consequences." *International Organization* 38, no. 1: 1–40.

Davis, Ralph. 1962. *The Rise of the British Shipping Industry in the Seventeenth and Eighteenth Centuries.* London: Macmillan.

———. 1975. *English Merchant Shipping and Anglo-Dutch Rivalry in the Seventeenth Century.* London: HMSO.

Deane, Phyllis. 1965. *The First Industrial Revolution.* Cambridge: Cambridge University Press.

Drewry Shipping Consultants. 1978. "Trends in Japanese Dry Bulk Shipping and Trade." London: Drewry Shipping Consultants.

Eckes, Alfred E., Jr. 1979. *The United States and the Global Struggle for Minerals.* Austin: University of Texas Press.

Elbaum, Bernard, and Frank Wilkinson. 1979. "Industrial Relations and Uneven Development: A Comparative Study of the American and British Steel Industries." *Cambridge Journal of Economics* 3, no. 3: 275–303.

Ellis, Joseph J. 1997. *American Sphinx: The Character of Thomas Jefferson.* New York: Knopf.

———. 2000. *Founding Brothers: The Revolutionary Generation.* New York: Vintage.

Encyclopedia Americana. 1995. Danbury, CT: Grolier.

Ewart, W. D., and H. Fullard, eds. 1972. *World Atlas of Shipping.* New York: St. Martin's Press.

Fearnleys. Various years. *Fearnleys Review.* Oslo: Fearnleys.

———. Various years. *World Bulk Trades.* Oslo: Fearnleys.

Fletcher, Max E. 1958. "The Suez Canal and World Shipping: 1869–1914." *Journal of Economic History* 18, no. 4: 556–73.

Gao, Bai. 2001. *Japan's Economic Dilemma: The Institutional Origins of Prosperity and Stagnation.* Cambridge: Cambridge University Press.

Garrod, P., and W. Miklius. 1985. "The Optimal Ship Size: A Comment." *Journal of Transport Economics and Policy* 19, no. 1: 83–89.

Girvan, Norman. 1980. "Economic Nationalism vs. Multinational Corporations: Revolutionary or Evolutionary Change?" In *Trilateralism: The Tri-Lateral Commission and Elite Planning for World Management,* ed. Holly Sklar, 437–67. Boston: South End Press.

Goldstone, Jack. Forthcoming. *The Happy Chance: The Rise of the West in Global Context, 1500–1850.*

Gorski, Philip S. 1993. "The Protestant Ethic Revisited: Disciplinary Revolution in Holland and Prussia." *American Journal of Sociology* 99, no. 2: 265–316.

———. 1995. "The Protestant Ethic and the Spirit of Bureaucracy." *American Sociological Review* 60, no. 5: 783–86.

Goto, S. 1984. *Japan's Shipping Policy.* Tokyo: Japan Maritime Research Institute.

Hammond World Atlas. 1971. Maplewood, NJ: Hammond Publications.

Harris, J. R. 1988. *The British Iron Industry, 1700–1850.* London: Macmillan.

Harvey, David. 1983. *Limits to Capital.* Chicago: University of Chicago Press.

Headrick, Daniel R. 1981. *The Tools of Empire: Technology and European Imperialism in the Nineteenth Century.* New York: Oxford University Press.

Hemming, John. 1978. *Red Gold: The Conquest of the Brazilian Indians.* Cambridge: Harvard University Press.

Hirschman, Albert. 1958. *The Strategy of Economic Development.* New Haven: Yale University Press.

Hobsbawm, Eric J. 1968. *Industry and Empire.* London: Penguin.

Hogan, William. 1971. *Economic History of the Iron and Steel Industry in the United States.* Vols. 1–4. Lexington, MA: Lexington Books.

———. 1991. *Global Steel in the 1990s: Growth or Decline?* Lexington, MA: Lexington Books.

Hugill, Peter. 1994. *World Trade Since 1431.* Baltimore: Johns Hopkins University Press.

IISI. International Iron and Steel Institute. www.worldsteel.org.

Innis, Harold A. 1933. *Problems of Staple Production in Canada.* Toronto: Ryerson Press.

———. 1956. *Problems in Canadian Economic History.* Toronto: University of Toronto Press.

Isard, Walter. 1948. "Some Locational Factors in the Iron and Steel Industry since the Early Nineteenth Century." *The Journal of Political Economy* 65, no. 3: 203–17.

Israel, Jonathan I. 1989. *Dutch Primacy in World Trade, 1585–1740.* Oxford: Clarendon Press.

———. 1995. *The Dutch Republic: Its Rise, Greatness, and Fall, 1477–1806.* Oxford: Oxford University Press.

JAMRI [Japan Maritime Research Institute]. 1991. *Changes in the World's Shipbuilding Facilities for Large Size Vessels and Future Prospects thereof.* JAMRI Report no. 40. Tokyo: JAMRI.

Jansson, J., and D. Shneerson. 1982. "The Optimal Ship Size." *Journal of Transport Economics and Policy* 16, no. 3: 217–38.

Jevons, William Stanley. 2001. "Of the Economy of Fuel." *Organization and Environment* 14, no. 1: 99–104.

JISF. Japan Iron and Steel Federation. www.jisf.or.jp.

Jörnmark, Jan. 1993. *Coal and Steel in Western Europe, 1945–1993: Innovative Change and Institutional Adaptation.* Gotebörg: University of Gotebörg Press.

Karl, Terry Lynn. 1997. *The Paradox of Plenty: Oil Booms and Petro-States.* Berkeley: University of California Press.

Katzman, Martin T. 1987. "Review Article: Ecology, Natural Resources, and Economic Growth: Underdeveloping the Amazon." *Economic Development and Cultural Change* 35, no. 2: 425–37.

Kendall, P. 1972. "A Theory of Optimum Ship Size." *Journal of Transport Economics and Policy* 6, no. 1: 128–46.

Kriedte, Peter. 1983. *Peasants, Landlords, and Merchant Capitalists.* Cambridge: Cambridge University Press.

Landes, David S. 1969. *The Unbound Prometheus: Technological Change and Industrial Development in Western Europe from 1750 to the Present.* Cambridge: Cambridge University Press.

Lash, Scott, and John Urry. 1987. *The End of Organized Capitalism.* Cambridge: Polity Press.

Latham, A. J. H. 1978. *The International Economy and the Underdeveloped World, 1865–1914.* London: Croom Helm.

Leitner, Jonathan. 1998. "Upper Michigan's Copper Country and the Political Ecology of Copper, 1840's–1930's." Ph.D. dissertation. University of Wisconsin–Madison.

Linebaugh, Peter. 1992. *The London Hanged: Crime and Civil Society in the Eighteenth Century.* Cambridge: Cambridge University Press.

Lower, A. R. M. 1973. *Great Britain's Woodyard: British America and the Timber Trade, 1763–1867.* Montreal: McGill-Queens University Press.

Maddison, Angus. 2002. *The World Economy: A Millennial Perspective.* Paris: Development Centre of the Organisation for Economic Co-operation and Development.

Mandel, Ernest. 1975. *Late Capitalism.* London: New Left Books.

Manners, Gerald. 1971. *The Changing World Market for Iron Ore, 1950–1980: An Economic Geography.* Baltimore: Johns Hopkins Press.

Marshall, Chris, ed. 1995. *Encyclopedia of Ships.* New York: Barnes and Noble Books.

Marshall, Jonathan. 1995. *To Have and Have Not: Southeast Asian Raw Materials and the Origins of the Pacific War.* Berkeley: University of California Press.

Marx, Karl. 1867. *Capital: A Critique of Political Economy.* Vol. 1. New York: International Publishers, 1962.

———. 1894. *Capital: A Critique of Political Economy.* Vol. 3. New York: International Publishers, 1967.

Mathias, P. 1969. *The First Industrial Nation: An Economic History of Britain, 1700–1914.* New York: Scribner's.

McMichael, P. 1984. *Settlers and the Agrarian Question: Foundations of Capitalism in Colonial Australia.* Cambridge: Cambridge University Press.

———. 1990. "Incorporating Comparison within a World-Historical Perspective: An Alternative Comparative Approach." *American Sociological Review* 55, no. 3: 385–97.

———. 1992. "Rethinking Comparative-Analysis in a Post-Developmentalist Context." *International Social Science Journal* 44, no. 3: 351–65.

Meinig, D. W. 1986. *Atlantic America, 1492–1800.* Vol. 1 of *The Shaping of America: A Geographical Perspective on 500 Years of History.* New Haven: Yale University Press.

———. 1993. *Continental America, 1800–1867.* Vol. 2 of *The Shaping of America.*

———. 1998. *Transcontinental America, 1850–1915.* Vol. 3 of *The Shaping of America.*

Merchant, Carolyn. 1983. *The Death of Nature: Women, Ecology, and the Scientific Revolution.* San Francisco: Harper and Row.

Mintz, Sidney W. 1986. *Sweetness and Power: The Place of Sugar in Modern History.* New York: Viking.

Misa, Thomas J. 1995. *A Nation of Steel: The Making of Modern America, 1865–1925.* Baltimore: Johns Hopkins University Press.

Misra, Joya, and Terry Boswell. n.d. "Hegemony and the Capitalism during the Age of Mercantilism: A Comparative Historical Analysis of Global Leading Sectors, 1550–1749." Unpublished manuscript.

Mitchell, B. R. 1998. *International Historical Statistics: The Americas, 1750–1993.* London: Macmillan Reference.

Modelski, George, and William Thompson. 1996. *Leading Sectors and World Powers: The Coevolution of Global Politics and Economics.* Columbia, SC: University of South Carolina Press.

Mokyr, Joel. 1990. *The Lever of Riches: Technological Creativity and Economic Progress.* New York: Oxford University Press.

Morison, Samuel Eliot. 1941. *The Maritime History of Massachusetts, 1783–1860.* Boston: Houghton Mifflin.

Murphy, R. Taggart. 1996. *The Weight of the Yen.* New York: Norton.

NISF. 1923. *Statistics of the Iron and Steel Industries.* London: National Iron and Steel Federation.

North, Douglass. 1958. "Ocean Freight Rates and Economic Development, 1750–1913." *Journal of Economic History* 18, no. 4: 537–55.

———. 1968. "Sources of Productivity Change in Ocean Shipping, 1600–1850." *Journal of Political Economy* 76, no. 5: 953–70.

NSC. Various years. *Annual Report.* Tokyo: Nippon Steel Corporation.

———. 1973. *History of Steel in Japan.* Tokyo: Nippon Steel Corporation.

O'Brien, Patricia A. 1992. "Industry Structure as a Competitive Advantage: The History of Japan's Post-war Steel Industry." *Business History* 34, no. 1: 128–59.

Ogura, Seiritsu, and Naoyuki Yoshino. 1988. "The Tax System and Fiscal Investment and Loan Program." In *Industrial Policy of Japan,* ed. Ryutaro Koyima, Masahiro Okuno, and Kotaro Suzumura, 121–54. Tokyo: Academic Press Japan.

O'Hearn, Denis. 2002. *The Atlantic Economy: Britain, the U.S., and Ireland.* Manchester: Manchester University Press.

Olson, Charles. 1947. *Call Me Ishmael: A Study of Melville.* San Francisco: City Lights Books.

Oman, Charles. 1984. *New Forms of International Investment in Developing Countries.* Paris: OECD Development Center.

Oman, Charles, et al. 1989. *New Forms of Investment in Developing Country Industries: Mining, Petrochemicals, Automobiles, Textiles, Food.* Paris: OECD Development Center.

Paige, Jeffery M. 1999. "Conjuncture, Comparison, and Conditional Theory in Macrosocial Inquiry." *American Journal of Sociology* 105, no. 5: 781–800.

Parker, William N. 1991. *America and the Wider World.* Vol. 2 of *Europe, America and the Wider World.* Cambridge: Cambridge University Press.

Parry, J. H. 1967. "Transport and Trade Routes." In *The Cambridge Economic History of Europe,* ed. E. E. Rich and C. H. Wilson, 4:155–222. Cambridge: Cambridge University Press.

Patrick, Hugh, and Hideo Sato. 1981. "The Political Economy of United States-Japan Trade in Steel." *Australia-Japan Research Centre Paper* No. 88. Canberra: Australia National University.

Peirce, Charles S. 1877. "The Fixation of Belief." *Popular Science Monthly* 12: 1–15.

Perlin, John. 1991. *Forest Journey: The Role of Wood in the Development of Civilization.* Cambridge: Harvard University Press.

Perlman, Jacob. 1934. "Railroads, Labor." In *Encyclopaedia of the Social Sciences,* 13:93–98. New York: Macmillan.

Perloff, Harvey, and Lowdon Wingo Jr. 1961. "Natural Resource Endowments and Regional Economic Growth." In *Natural Resources and Economic Growth,* ed. Joseph J. Spangler, 191–212. Washington DC: Resources for the Future.

Phillips, Carla Rahn. 1990. "The Growth and Composition of Trade in the Iberian Empires, 1450–1750." In *The Rise of Merchant Empires: Long Distance Trade in the Early Modern World, 1350–1750,* ed. James Tracy, 34–101. Cambridge: Cambridge University Press.

Priest, R. Tyler. 1996. "Strategies of Access: Manganese Ore and U.S. Relations with Brazil, 1894–1953." Ph.D. dissertation. University of Wisconsin–Madison.

Quiroz Norris, Alfonso W. 1983. "Las Actividades Comerciales y Financieras de la Casa Grace y la Guerra del Pacífico." *Historica* 8, no. 2: 214–54.

Ricardo, David. 1817. *On the Principles of Political Economy and Taxation.* London: J. Murray, 1983.

Robinson, William I. 2001. "Transnational Processes, Development Studies, and Changing Social Hierarchies in the World System: A Central American Case Study." Paper presented at ISA meetings, Chicago, Illinois.

Rosenberg, N., and L. Birdzell. 1986. *How the West Grew Rich: The Economic Transformation of the Industrial World.* New York: Basic Books.

Ross, Eric. 1978. "The Evolution of the Amazonian Peasantry." *Journal of Latin American Studies* 10, no. 2: 193–218.

Rostow, Walt. 1960. *The Stages of Economic Growth: A Non-Communist Manifesto.* Cambridge: Cambridge University Press.

Santos, Roberto. 1968. "O Equilibrio da firma 'aviadora' e a significação econômico-institucional do 'aviamento.'" *Pará Desenvolvimento* 3. Belem: IDESP.

———. 1980. *Historia Economica da Amazonia, 1800–1920.* São Paulo: T. A. Queiroz.

Schumpeter, Joseph. 1934. *The Theory of Economic Development: An Inquiry into Profits, Capital, Credit, Interest, and the Business Cycle.* New York: Oxford University Press.

See, Henri. 1928. *Modern Capitalism: Its Origin and Evolution.* Trans. by H. R. Vanderblue. New York: Adelphi Company.

Shafer, D. Michael. 1994. *Winners and Losers: How Sectors Shape the Developmental Prospects of States.* Ithaca: Cornell University Press.

Shepherd, James F., and Gary M. Walton. 1972. *Shipping, Maritime Trade, and the Economic Development of Colonial North America*. Cambridge: Cambridge University Press.

Sklair, Leslie. 2000. *The Transnational Capitalist Class*. Oxford: Blackwell.

Staley, Eugene. 1937. *Raw Materials in Peace and War*. New York: Council on Foreign Relations.

Stopford, M. 1988. *Maritime Economics*. London: Unwin Hyman.

Swank, James. 1892. *History of the Manufacture of Iron in All Ages*. New York: Burt Franklin, 1965.

Sweet, David. 1974. "A Rich Realm of Nature Destroyed." Ph.D. dissertation. University of Wisconsin–Madison.

Temin, Peter. 1964. *Iron and Steel in Nineteenth-Century America: An Economic Inquiry*. Cambridge: MIT Press.

Tex Report. 1994a. *Coal Manual 1994*. Tokyo: The Tex Report Company Ltd.

———. 1994b. *Iron Ore Manual 1993–94*. Tokyo: The Tex Report Company Ltd.

Thompson, Mark. 1994. *Queen of the Lakes*. Detroit: Wayne State University Press.

Tilly, Charles. 1995a. "Macrosociology, Past and Future." *Newsletter of the Comparative and Historical Sociology Section of the American Sociological Association* 8, nos. 1/2: 3–4.

———. 1995b. "To Explain Political Processes." *American Journal of Sociology* 6: 1594–1610.

Todd, D. 1991. *Industrial Dislocation: The Case of Global Shipbuilding*. London: Routledge.

Tomich, Dale. 1994. "Small Islands and Huge Comparisons: Caribbean Plantations, Historical Unevenness, and Capitalist Modernity." *Social Science History* 18, no. 3, 339–58.

———. 2004. "Atlantic History and World Economy: Concepts and Constructions." *Protosociology* 20: 102–23.

UNCTAD [United Nations Conference on Trade and Development]. Various years. *Review of Maritime Transport*. New York: United Nations.

Unger, Richard. 1978. *Dutch Shipbuilding before 1800: Ships and Guilds*. Amsterdam: Van Gorcum.

USBM. 1987. "The Minerals Industries of Japan." *Minerals Yearbook*, 499–528. Washington, DC: United States Bureau of Mines.

U.S. Department of Commerce. 1925. *Plantation Rubber in the Middle East*. Washington, DC: GPO.

U.S. Geological Survey and U.S. Bureau of Mines. Various years. *Mineral Resources of the United States*. Washington, DC: GPO.

Van Houtte, J. A. 1977. *An Economic History of the Low Countries, 800–1800*. London: Weidenfeld and Nicolson.

Vance, James E., Jr. 1990. *Capturing the Horizon: The Historical Geography of Transport since the Transportation Revolution of the Sixteenth Century*. Baltimore: Johns Hopkins University Press.

Von Thünen, Johann. 1966. *Isolated State*. Oxford: Pergamon Press.

Wallerstein, I. 1980. *The Modern World-System II: Mercantilism and the Consolidation of the European World-Economy, 1600–1750*. New York: Academic Press.

———. 1982. "Dutch Hegemony in the Seventeenth-Century World-Economy." In *Dutch Capitalism and World Capitalism*, ed. Maurice Aymard, 93–146. Cambridge: Cambridge University Press.

Warren, Kenneth. 1973. *The American Steel Industry: A Geographical Interpretation.* Oxford: Clarendon Press.

———. 1975. *World Steel: An Economic Geography.* New York: David and Charles Newton Abbot/Crane, Russak.

Weber, Max. 1904–5. *The Protestant Ethic and the Spirit of Capitalism.* New York: Scribner, 1958.

Weinstein, Barbara. 1983. *The Amazon Rubber Boom, 1850–1920.* Stanford, CA: Stanford University Press.

Williams, Judith Blow. 1972. *British Commercial Policy and Trade Expansion, 1750–1850.* Oxford: Clarendon Press.

Willis, Susan. 1991. *A Primer for Daily Life.* New York: Routledge.

Wilson, Charles. 1973. "Transport as a Factor in the History of Economic Development." *Journal of European Economic History* 2, no. 2: 320–37.

Wittfogel, Karl. 1929. "Geopolitics, Geographical Materialism, and Marxism." *Antipode* 17, no. 1 (1985): 21–72.

Woo, Jung-en. 1991. *Race to the Swift: State and Finance in Korean Industrialization.* New York: Columbia University Press.

Wright, Gavin. 1990. "The Origins of American Industrial Success, 1879–1940." *American Economic Review* 80, no. 4: 651–68.

Yamamoto, S. 1980. "Shipbuilding." In *An Industrial Geography of Japan,* ed. Kiyoji Murata, 163–75. New York: St. Martin's Press.

Yonezawa, Yoshie. 1988. "The Shipbuilding Industry." In *Industrial Policy of Japan,* ed. Ryuaro Komiya, Masahiro Okuno, and Kotaro Suzumura, 445–49. Tokyo: Academic Press.

Zakaria, Fareed. 1998. *From Wealth to Power: The Unusual Origins of America's World Role.* Princeton: Princeton University Press.

Zimmerman, Warren. 2002. *First Great Triumph: How Five Americans Made Their Country a World Power.* New York: Farrar Straus Giroux.

Index